CW01646226

Safe Moving of Objects & People

Made Easy

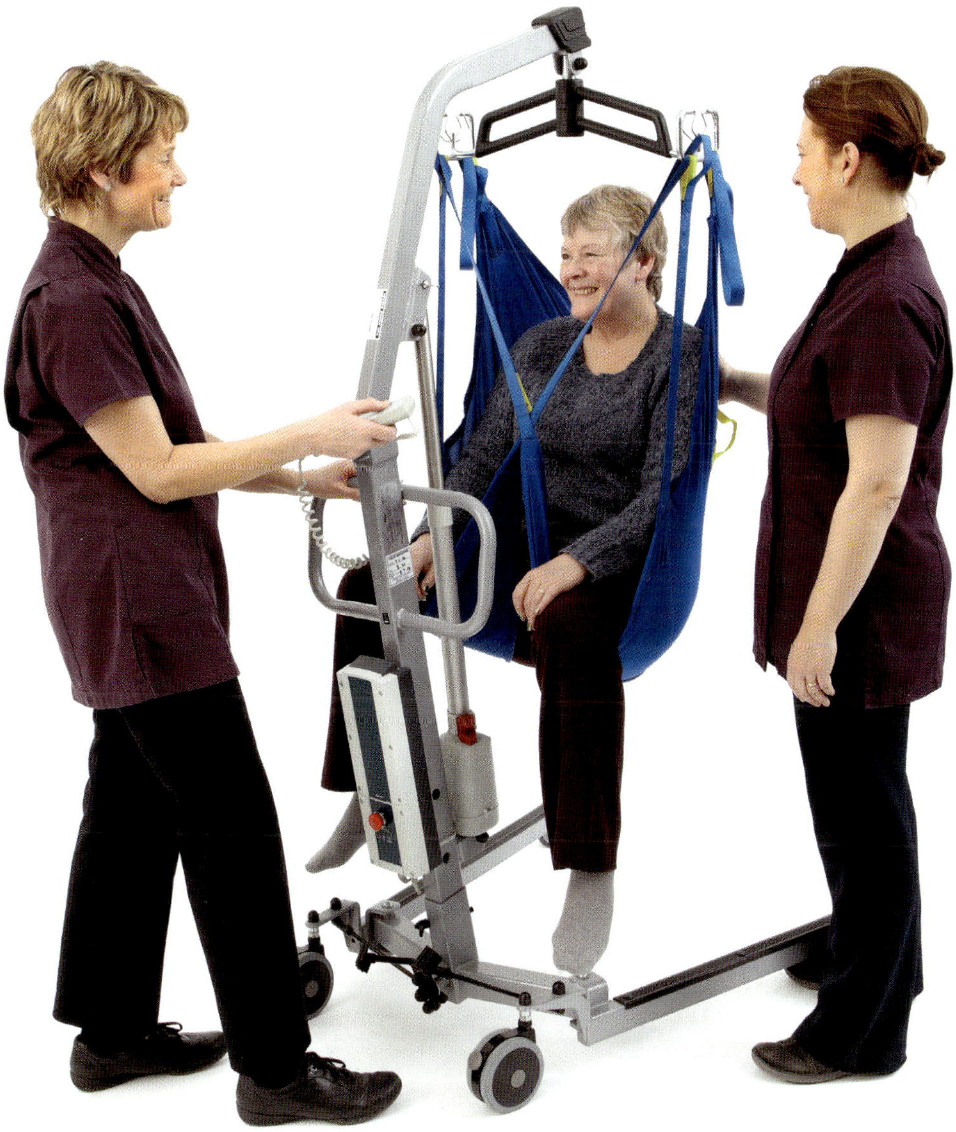

Edition 3.1

Publisher: Qualsafe.com

This book is designed as a learning guide to a safe moving of objects and people course and cannot replace hands on training in the skills of moving objects and people safely. Whilst every effort has been made to ensure the accuracy of the information contained within this book, the publisher does not accept liability for any inaccuracies or for any subsequent injuries sustained, however caused.

Safe Moving of Objects & People

Made Easy

Contents

Introduction to Manual Handling

Definition of Manual Handling

Manual handling is any activity that includes lifting or putting down, pulling or pushing, carrying, supporting or moving an object, person or animal using the hand or any other part of the body. Many people make the assumption that manual handling only applies when items are lifted, but it could be ANY movement of an object.

Definition of Moving and Handling People

'Moving and handling people' is the term often used to refer to the handling of a person to help their own movement or to enable movement when they are unable to move independently.

Common Causes of Manual Handling Workplace Injuries

The most common causes of manual handling injuries are:

- Poor posture
- Heavy lifting or using excessive force
- Excessive repetition of a task
- Lack of time to recover from a task
- Twisting and stooping activities
- Pushing or pulling loads over a prolonged distance
- Reaching or lifting away from the body
- Standing or sitting in a static position without regular changes in posture
- Person's own susceptibility to injury

About a third of all workplace accidents reported to the Health and Safety Executive *(HSE)* are as a direct result of handling, lifting and carrying.

The Cost of Manual Handling Injuries

Manual handling injuries can have a serious effect on both employers and employees.

Business costs

- Legal advice and insurance costs
- Legal action, possibly including compensation
- Sick pay
- Lost production and sales
- Equipment and stock losses or damages
- Time and money spent on investigation and improvements
- Recruitment and payment of cover/ replacement staff

However, these can fade into insignificance when weighed against the human costs.

Human costs

- Pain and suffering
- Disability, possibly long-term
- Health care and rehabilitation
- Loss of earnings
- Emotional and financial burden on families
- Potential loss of life

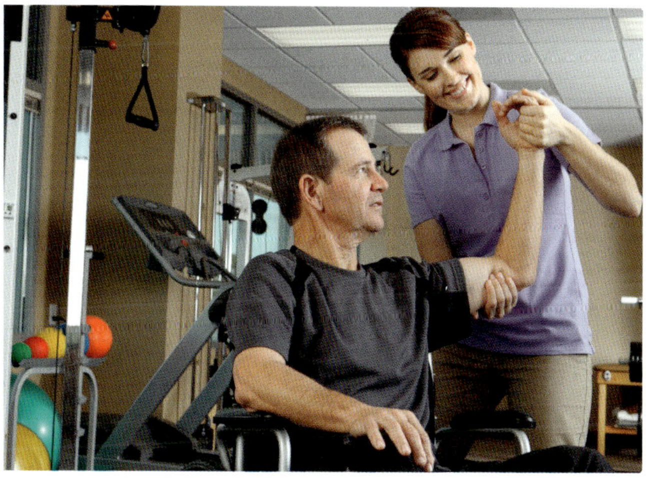

Risk Controls

When you weigh up the human and business costs that manual handling injuries cause, it becomes apparent that avoidance of the injury in the first place is to everyone's advantage.

The employer and employee need to consider:

The Hazard – things that have potential to do harm

The Risk – the likelihood of the hazard actually doing harm

Control Measures – actions or procedures put into place to reduce risk to an acceptable level

Manual Handling and the Law

There are a number of laws relating to manual handling that cover the responsibilities of employers, employees and the self-employed.

Health and Safety at Work etc. Act *(1974)*

The Health and Safety at Work etc. Act came in to force in 1974 and covers everyone at work *(apart from persons attending domestic premises)*. The Act makes it clear that everyone has a part to play in health and safety and aims to encourage high standards and prevent people coming to harm while in their working environment. The responsibilities of various parties under the Act are detailed below:

Employers

The Act places a 'duty of care' on employers to introduce good health and safety practices and provides a framework for health and safety regulations. They must ensure the health, safety and welfare of employees, visitors and the public by providing:

- A written Health and Safety Policy *(if they employ 5 or more people)*
- Safe equipment and machinery
- Safe systems of work
- Information, instruction, supervision and training
- A safe working environment
- Adequate welfare facilities

Employees

Under the Health and Safety at Work etc. Act all employees have a duty to:

- Act in a way that does not put themselves or others at risk
- Co-operate with their employer on any health and safety matters

Self-employed

Health and safety law will apply to the self-employed if:

- They employ other people
- Their work presents a potential risk to the health and safety of others
- Their work activity is specifically mentioned in the Health and Safety at Work etc. Act 1974 (General Duties of Self-Employed Persons) (Prescribed Undertakings) Regulations 2015

When relevant, self-employed people share many of the same duties as employers. They must make sure their work does not endanger themselves or others and inform people of hazards, e.g. by putting up signs to alert others to hazards.

Occupiers of Buildings

Persons in control of buildings have duties to people who are not employed by them but work at or visit the premises, e.g. customers, contractors, visitors, etc.

Designers, Manufacturers and Installers

Equipment and articles must be designed, constructed and tested to ensure they will work safely and any substances used in their manufacture must be safe. Adequate information on the safe use of the equipment must be provided.

Manual Handling Operations Regulations *(1992)*

This regulation is specific to manual handling responsibilities and outlines a clear "Hierarchy of Control" – AVOID, ASSESS, REDUCE, REVIEW *(for more details on "Hierarchy of Control" see Risk Assessment chapter on page 16).*

Employer's responsibilities

When possible, employers should **AVOID** the need for employees to perform any manual handling tasks at work which could result in them being injured.

Employers are also responsible for contracted and volunteer staff.

When it is not possible to avoid manual handling tasks at work they must:

- Make a suitable **ASSESS**ment of the task to be undertaken
- Take steps to **REDUCE** the risk of injury
- Inform employees of the weight of the load
- **REVIEW** procedures if there has been a significant change to the manual handling operation or task

Employee's responsibilities

All employees have a duty to:

- Follow safe systems provided

Management of Health and Safety at Work Regulations *(1999)*

Employer's responsibilities

Besides carrying out a risk assessment, employers also need to:

- Make arrangements for implementing the health and safety measures identified as necessary by the risk assessment
- Appoint a competent person *(someone who has sufficient training, experience or knowledge to enable them to assist the employer in complying with Health and Safety legislation)*
- Set up emergency procedures
- Provide clear information and training to employees
- Pay particular attention to high risk groups such as under 18s and new or expectant mothers
- Work together with other employers sharing the same workplace

Employee's responsibilities

- Use any equipment and safety device given to them by their employer in the way in which they were trained
- Advise an employer if they notice any risk or danger in the workplace

Reporting of Injuries, Diseases and Dangerous Occurrences Regulations *(2013)* – *(RIDDOR)*

RIDDOR requires employers *(including the self-employed)* or those in control of premises to report any work-related deaths, specified injuries, occupational diseases, reportable dangerous occurrences and non-fatal accidents requiring hospital treatment to non-workers to the HSE.

Injuries that occur at work that mean the worker is absent or unable to do their work for more than 7 days need to be reported within 15 days of the accident occurring.

By law, employers must also keep a record of all injuries that result in workers being unable to work for more than 3 days *(an accident book entry is acceptable).*

However, it is good practice to record all accidents and near misses.

Human Rights Act *(1998)*

The Human Rights Act *(1998)* means that the rights from the ECHR *(European Convention of Human Rights)* are now enforceable in our own courts.

The Act defines the basic rights and freedoms of people in the UK. The rights and freedoms that could be specifically related to the safe moving and handling of people are:

- Freedom from torture and inhuman or degrading treatment
- Right to respect for your private and family life

Mental Capacity Act *(2005)*

This Act was put in place to protect people who are unable to make their own decisions. It allows people to plan ahead for a time when they may lose their mental capacity as they can appoint a person in advance to make decisions about their welfare or property for them. When future plans have not been put in place, this Act makes sure all decisions made about a person are made in their best interest or with the best intent.

Care Act *(2014)*

The Care Act (2014) is applicable to people who require care and support and those who provide care for another person, usually a relative or friend, in their day-to-day life. It is not applicable to those who care for someone through a professional or voluntary organisation. The Act was put in place to consolidate numerous other care laws in England and to provide consistency in the way people receive care and support. The Act highlights local authorities' responsibility to assess the care needs of people living in their area, provide information and advice on care options and supply appropriate services to meet people's needs and preferences. For the first time, the support needs of those providing care will be assessed alongside those they care for. The Act also provides a clear legal framework for how local authorities and other parts of the health care system should safeguard adults at risk from abuse or neglect. It has also reformed the system for funding and introduced national rules on how local authorities determine who is eligible to receive care and support.

Provision and Use of Work Equipment Regulations *(1998)* – *(PUWER)*

and

Lifting Operations and Lifting Equipment Regulations *(1998)* – *(LOLER)*

These regulations specifically cover equipment and are discussed in more detail on pages 27 and 34.

The courts can imprison those found guilty for up to 2 years.

Penalties and Enforcement

The HSE and Local Authority Environmental Health Departments are responsible for enforcing most health and safety law at work. In cases where advice given has not been followed or they feel there is an unacceptable level of risk, legal notices can be served. Failure to comply with a notice is a criminal offence. Substantial fines can be issued and in severe cases, the courts can imprison those found guilty for up to 2 years.

The HSE's job is to protect people from risk and subsequent injury at work. Guidance on all health and safety related topics can be found on their website: www.hse.gov.uk.

The Musculoskeletal System

The musculoskeletal system is made up of two systems that work together to allow you to support and move your own body – the skeletal system *(bones and joints)* and the skeletal muscle system.

Safe manual handling should protect your whole body, but specific emphasis is put on protecting the spine as this is the most commonly injured area.

The Spine

The spine has 3 main functions:

1. **To protect the spinal cord**
2. **To allow movement**
3. **To support the upper body**

There are 33 bones in the spine. These bones are called vertebrae.

Ligaments hold the vertebrae together. Between each individual bone there is a disc – these discs act as shock absorbers during activities such as running, jumping and walking. The outer part of the disc is made of lots of layers of cartilage and the inner nucleus is a softer consistency. The discs enable the spine to flex and extend, allow smooth movement of the spine and help bond the vertebrae together.

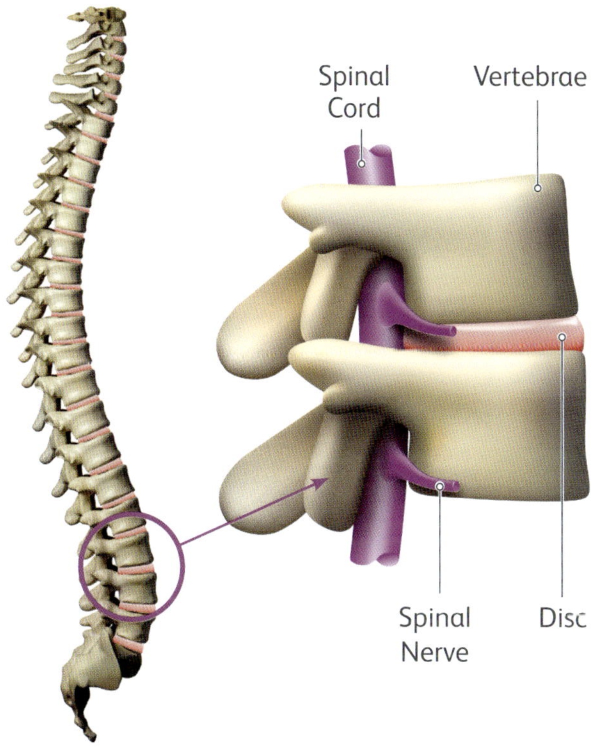

Normal Disc

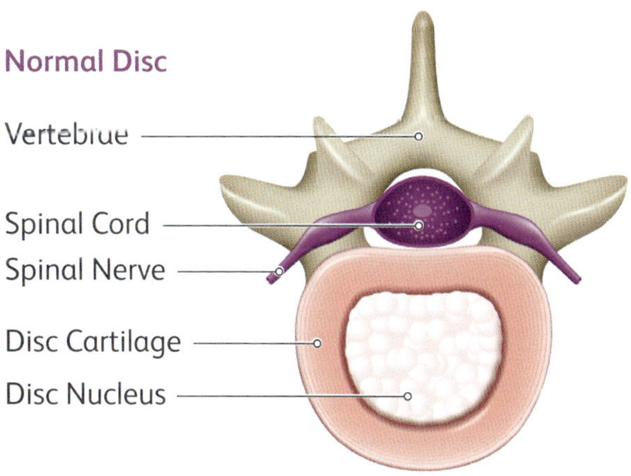

Injured Disc

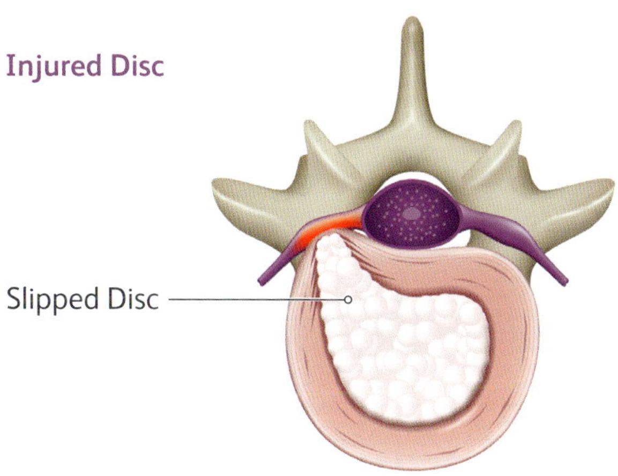

A common injury to the spine is a slipped disc. This is where the layers of cartilage of the disc can become cracked and the inner nucleus pushes out of the crack and presses on the spinal nerve *(see diagram)*.

The shape of the vertebrae creates a canal. The spinal cord runs down this canal from the base of the brain. Therefore the spine plays a significant role in protecting the spinal cord. Nerves from the spinal cord branch out and leave the spine through spaces between the vertebrae.

Posture

When viewed from the side, the spine has three natural curves. It is important to maintain these curves to protect the spine and optimise function. If the posture becomes excessive it is described as "poor posture" *(see diagram below)*.

As a population we have high levels of spinal problems. Ongoing "poor posture" is thought to be a large cause of these spinal problems. Even if there is no manual handling involved, poor posture can cause cumulative strain on the discs and muscles.

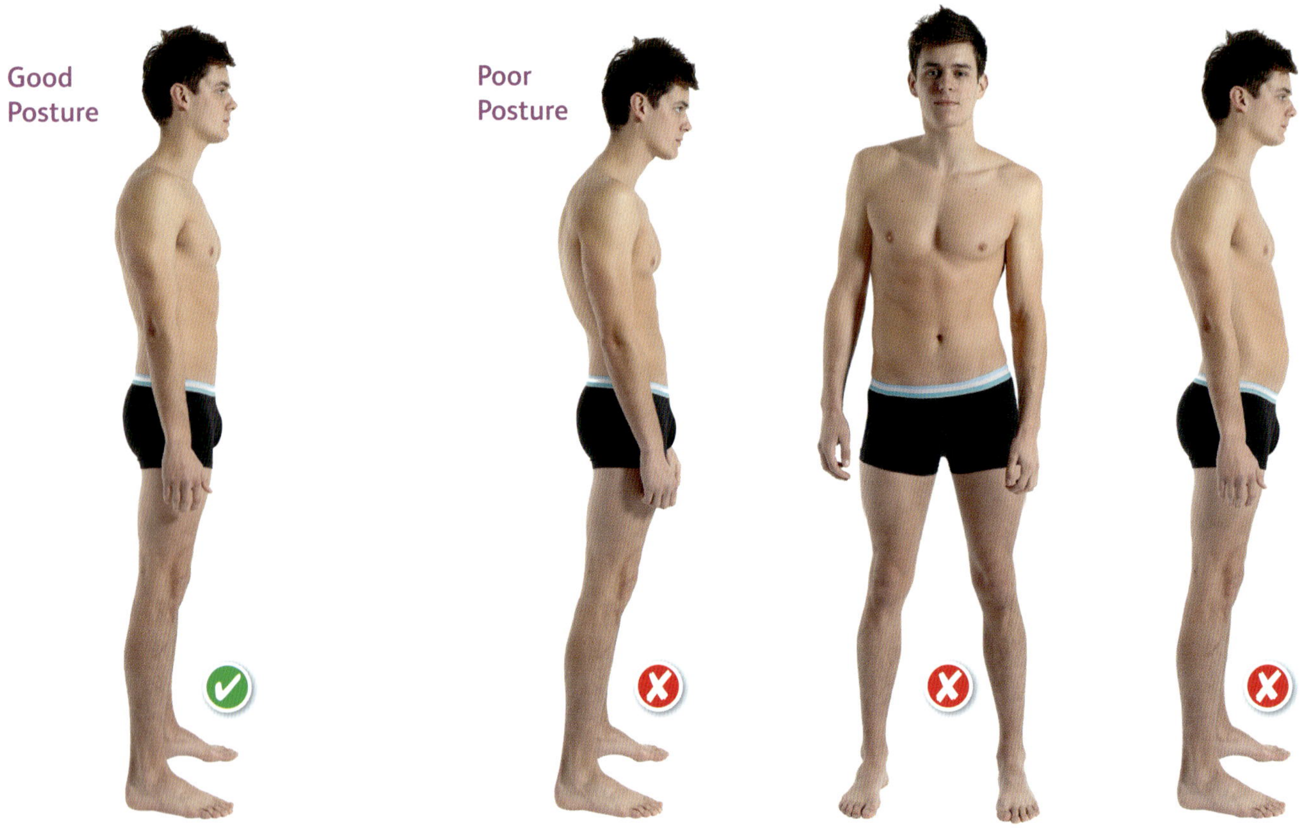

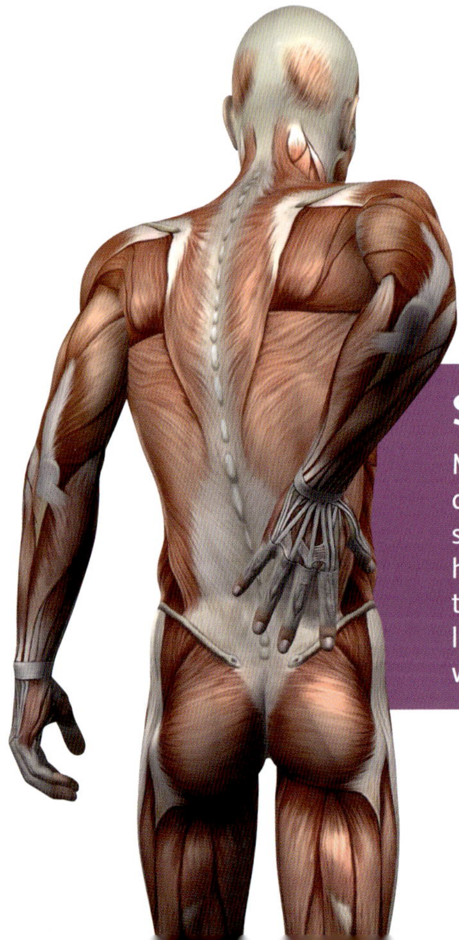

Spinal Muscles

Many muscles and tendons help to provide spinal balance, stability and mobility. Muscles are attached to the vertebrae by tendons. Each separate back muscle is comparable to the size of the muscles in the hand that control the fingers, therefore it is important to note that they are not designed to do strong prolonged contractions such as lifting. Static work also tires these muscles quickly due to their size which can make you more prone to injury.

Mechanics of Movement

Understanding your own body mechanics plays a crucial role in safe manual handling.

Centre of Gravity

Every person and item has a centre of gravity. The centre of gravity is the point at which the whole weight of an object can be considered to act and therefore at which all parts of an object are balanced. When a person is standing still, their centre of gravity is said to be located inside their body at about the level of their navel *(belly button)*. The position of the centre of gravity will move if you change your body position.

The centre of gravity can move with activity of the arms or body. The further away from the centre of gravity the activity is, the greater the pressure in the spine.

The Lever System

Your body's musculoskeletal system uses levers to produce movement so it is important to understand the basics about how levers work. All levers require a fulcrum *(pivot point)* to function. As demonstrated in the diagrams below, changing the distance between the fulcrum and the load can greatly affect the amount of force required to balance a load.

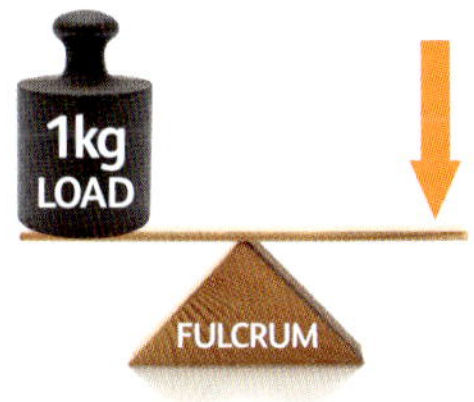

Figure 1

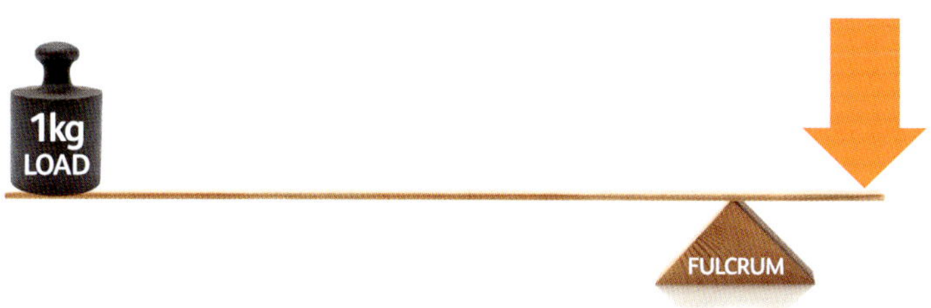

Figure 2

In **figure 1** there are equal distances from the load to the fulcrum and from the fulcrum to the counterbalancing force. The counterbalancing force must be equal to the weight of the load to keep the lever balanced.

In **figure 2** the weight of the load and the distance between the fulcrum and the counterbalancing force is unchanged. The distance between the load and fulcrum however has increased by 5 times. Because of this extra distance, the counterbalancing force must be 5 times greater to keep the lever balanced.

Actually lifting a load requires a force greater than that which is needed to counterbalance it. When you lean over *(instead of bending your knees to lift an object)* you are increasing the amount of pressure on your back. Additional pressure is put on your back muscles as they are counterbalancing the weight of both the object and your upper body.

A longer lever increases the pressure on your back.

A shorter lever puts less pressure on your back.

Body Positioning

The stability of your body when performing manual handling tasks plays an important role. You should always make sure you have a stable base when moving objects with the centre of gravity falling within your feet. Your feet should be positioned shoulder width apart to give you better balance and you should bend your knees, keep your chin up and try to maintain the normal curvature of your spine. This will enable the weight to be evenly distributed from the spine to the pelvis where your much stronger leg muscles can apply the force required to move the load. To decrease the pressure on your spine, you should hold the load close to your body. The further away from the centre of gravity the load is, the greater the pressure on the spine.

Feet Positioning

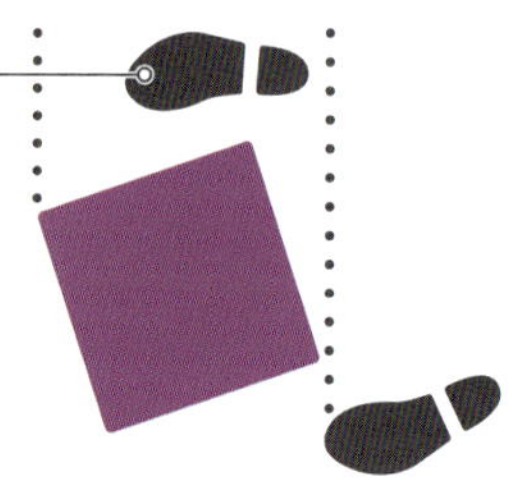

NOTE
Always make sure you have a stable base when moving objects.

Ergonomics

Ergonomics is the science of fitting the task to the person by designing equipment and apparatus that fit well with the human body and the way that it moves. People come in all shapes and sizes so when possible work stations should be designed with adjustable features, such as adjustable seats and work platforms, so that employees can work comfortably. Failure to do so can result in people stooping, stretching, overreaching and twisting. This can cause muscle fatigue, strain and possible long term damage or deformation.

Other factors to consider when risk assessing the ergonomics of work activities are:

- The degree of force needed to do the job
- The environmental conditions – lighting and temperature
- Adequacy of rest periods
- The repetitiveness of the task

Any machine expected to carry out repetitive motions should be regularly maintained with parts replaced as they wear out. Failure to maintain machinery may result in breakdown. People may also be expected to carry out similar repetitive tasks and suffer wear and tear as a result, causing them to suffer permanent damage and being unable to work. Any activity involving high numbers of repetitions should be looked at carefully and redesigned to avoid this as much as possible.

Remedies

- Alternating tasks and work breaks to avoid repetitive strain injuries
- Providing a comfortable working temperature and adequate ventilation
- Providing sufficient lighting and minimizing glare by glazing windows or supplying blinds
- Minimise vibration
- Mechanise the task
- Regular health tests may be necessary and should be promoted by employers
- Training will also help employees have a greater awareness of good practice

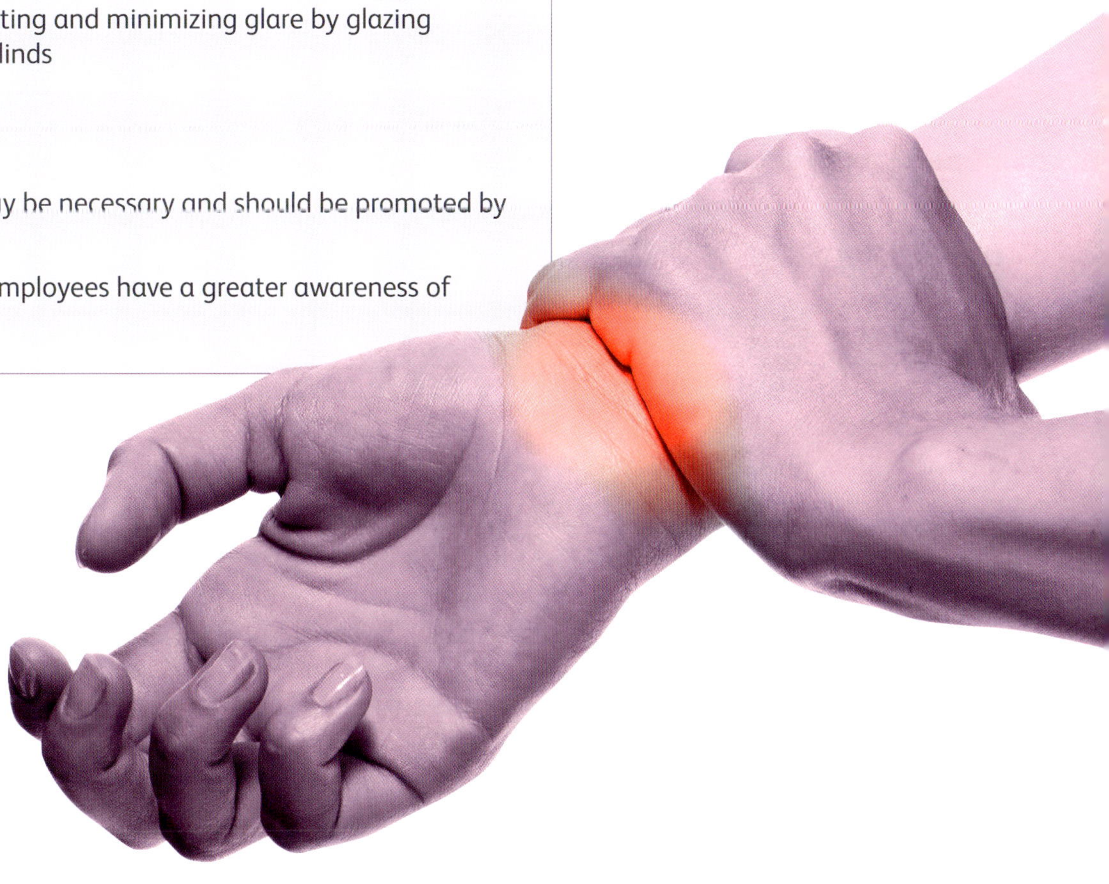

Types of Manual Handling Injuries

Manual handling injuries can occur as a result of:

- A single event or accident
- Repetitive tasks
- Poor posture
- Incorrect handling techniques

The vast majority of manual handling injuries reported to the HSE are as a result of an injury to the lower back, however, many other areas of the body can also be affected.

Some common types of manual handling injuries:

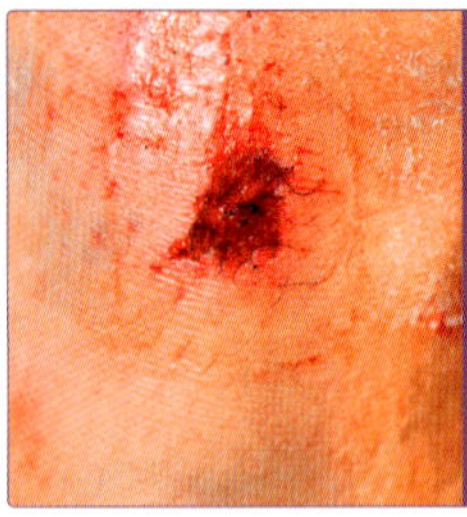

Abrasions, cuts and fractures

Broken skin and broken bones can occur in various parts of the body. These are often caused by a single event or accident.

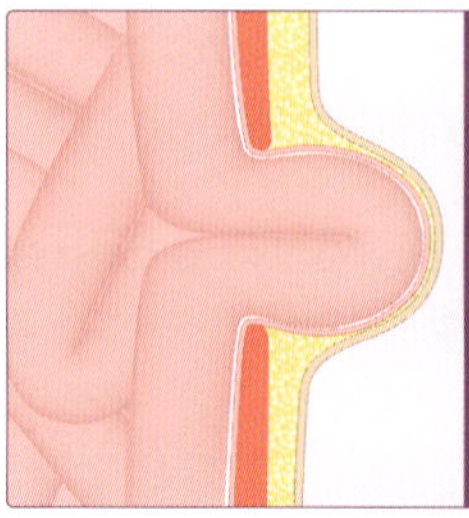

Hernia

A protrusion of tissue through a weakness in the muscle or surrounding tissue wall. This can be caused by lifting heavy objects incorrectly.

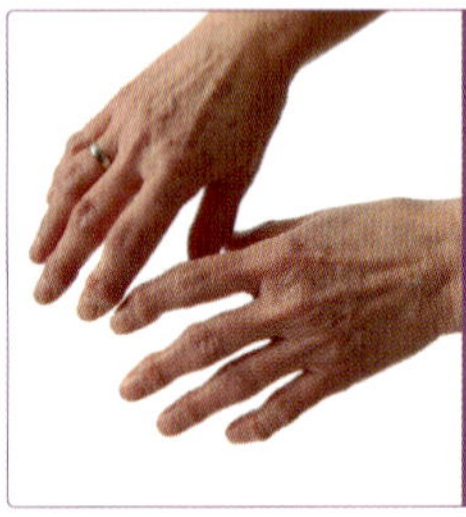

Arthritis

Inflammation of a joint. This can be caused by wear and tear due to repetitive tasks.

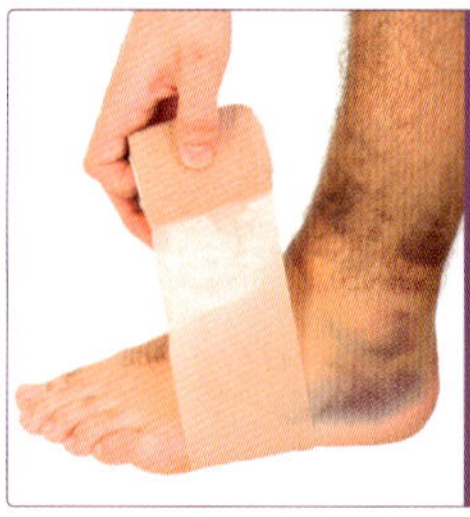

Muscle or ligament sprain or strain

The muscle or ligament fibres in joints tearing as a result of overstretching *(pulled muscle)*.

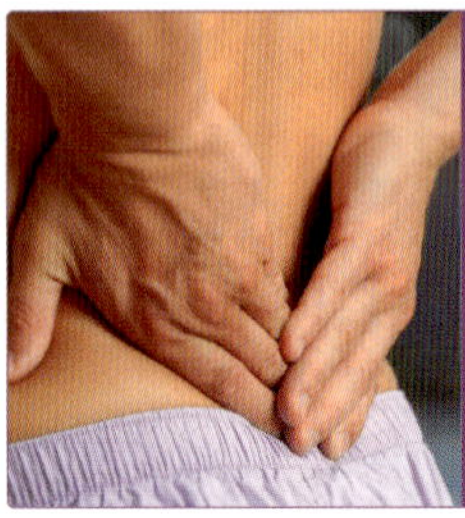

Back pain

Pain may be a result of muscular or nerve damage. Causes of back pain include poor static posture and heavy lifting.

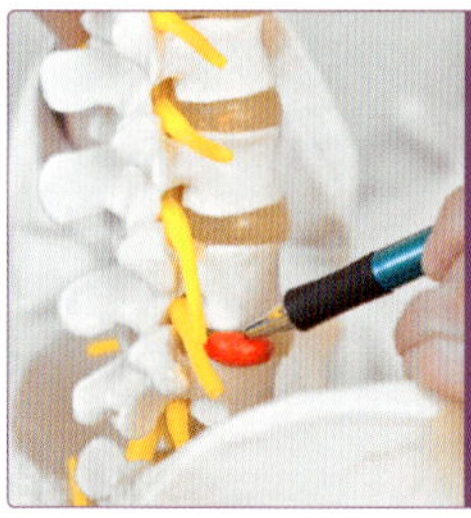

Slipped disc

Layers of cartilage of the disc can become cracked and the inner nucleus pushes out of the crack and presses on the spinal nerve *(see diagram on page 9)*.

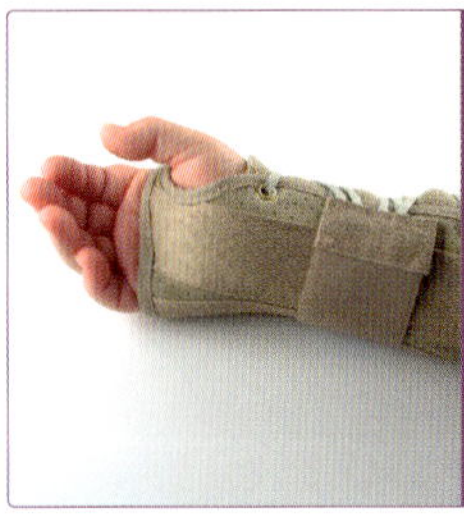

Carpal tunnel syndrome

A painful condition of the hands and fingers where a major nerve in the hand and wrist is compressed. Repetitive motions can cause this.

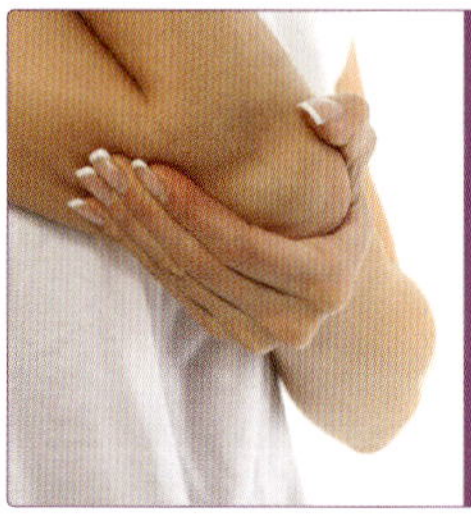

Tendonitis

Inflammation of a tendon caused by overuse.

No occupation is free from potential manual handling injuries, but some occupations are particularly susceptible to spine and lower back injuries, e.g. carers, nurses and delivery personnel.

Other occupations are more at risk of developing injuries as a result of repetitive tasks, e.g. hairdressers and factory line workers.

You must remember that manual handling injuries can happen in your everyday life too. Therefore the safe handling principles and good posture that you apply to your work life are also relevant to your life outside work.

Factors that Increase Risk of Harm

No individual is immune from manual handling injuries, but there are some factors that may make a person more susceptible to harm.

Task related factors that increase risk of injury:

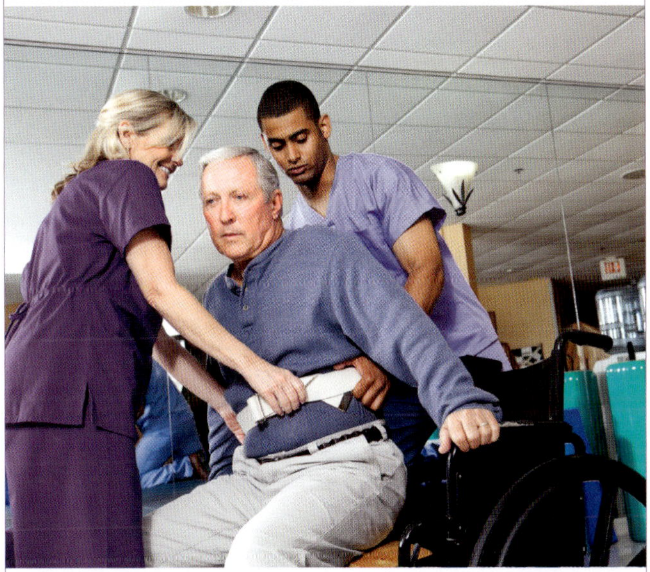

- Heavy manual labour
- Awkward postures
- Vibration
- Static work – any position held for longer than 30 seconds is considered to be a static posture
- Bending and twisting
- Carrying loads away from your body and away from your centre of gravity

Personal factors that increase risk of injury:

- Being overweight or unfit
- Previous or existing injuries
- Motherhood or pregnancy
- Age – risk of lower back disorders increases with the number of years you have worked
- Smoking
- Genetic factors

Risk Assessment

When examining the topic of risk assessment, it is useful to understand some of the terms commonly used:

Reasonably practicable

A risk/sacrifice computation made to see what safety measures are reasonable under given circumstances of cost/benefit

Suitable and sufficient

Is the risk assessment appropriate for the given circumstances?
Is it detailed enough with regard to the level of risk involved?

Competent person

A person who has sufficient training, experience, knowledge or other relevant qualities to enable them to undertake the task or job.

Hierarchy of control

Control measures to reduce risks listed in order of priority *(Avoid, Assess, Reduce and Review – as detailed on page 17)*

The Management of Health and Safety at Work Regulations *(1999)* requires the employer to arrange for a competent person*(s)* to carry out a detailed risk assessment. It should identify all workplace hazards, quantify the risks and introduce control measures to reduce the risk of injury or illness to the workforce. **If more than 5 people are employed, the assessment must be recorded**. Specific risk assessments should be completed for manual handling activities.

A risk assessment is simply a careful examination of what could cause harm to people so that you can consider whether you have taken enough precautions or should do more to prevent harm. In the case of safe moving and handling of people, both the carer and the person being moved need to be considered.

Risk assessments will help identify what could go wrong, how likely it is to happen and how serious the result could be. The employer then needs to put control measures in place to prevent the problem occurring.

Carers should also complete on the spot risk assessments before moving a person to make sure current written risk assessments are still appropriate on that specific day.

Four Steps of a Manual Handling Risk Assessment

1 AVOID

- Can the task be avoided?
- Could a piece of equipment perform the task instead of a person?

2 ASSESS

Where a manual handling task cannot be avoided, remember T.I.L.E.

T **Task** – does the task involve twisting, reaching forward, carrying loads for long distances or carrying loads away from the body? Does the task involve awkward posture, heavy manual labour or repetitive motions?

I **Individual** – do they need Personal Protective Equipment *(PPE)*? Have they received appropriate training?

L **Load** – consider the weight, size and stability of load. In the case of handling people, an estimated weight of the client should form part of the risk assessment and be communicated to staff as part of the client's personal handling plan *(see page 19)*

E **Environment** – such as uneven or slippery floors, poor lighting and unhelpful temperatures. Are there seasonal weather factors to consider?

See page 18 for more detail.

Decide on Who Might be Harmed and How

It is not just the person conducting the task that may be affected, but also anyone nearby. If you are moving a person using manual handling techniques, the person being moved should also be considered.

Factors specific to individuals should be considered when assessing risk, e.g. their level of fitness and any previous injuries. Separate risk assessments will need to be completed for employees who are more vulnerable such as pregnant workers or young inexperienced staff.

3 REDUCE

Where a manual handling task cannot be avoided and you need to reduce the level of risk, remember T.I.L.E. – ***see page 18 for more detail.***

Consider the consequences of injury or harm. Could someone be seriously injured or killed? Could a lot of people be affected? Is it likely to occur? If the answer is yes, then these hazards should be addressed as a priority and further controls put in place to reduce the risk to an acceptable level. Could lifting equipment be used to decrease risk? Could the way a task is organised reduce the risk of harm?

4 REVIEW

It is a legal requirement that businesses with 5 or more staff formally record their significant findings. It is important to implement the controls that have been identified, make sure that staff are trained in new procedures and to check that they are being followed correctly.

Reviews should take place when:

- Procedures are revised
- Workplace layout is reorganised
- New equipment is going to be used
- An accident or near miss occurs
- The law changes
- The Task, Individual, Load or Environment *(T.I.L.E.)* changes
- The client's needs change

Reviews should also take place on a regular basis *(possibly annually)* to make sure nothing has been missed or to identify new techniques, new equipment or scientific developments that could be introduced to improve safety. Arrangements should be in place to monitor manual handling activities to make sure the correct procedures, techniques and equipment are being used.

Control Measures and Reducing the Risk of Harm

Where a manual handling task **cannot be avoided** remember **T.I.L.E.** *(Task, Individual, Load, Environment).*

	You should consider:	The risk of injury can be reduced by:
T Task	• How will the task be undertaken? • Can mechanical lifting aids be used? • Will it involve stretching, twisting or bending? • Is there any repetitive handling? • Does the task involve holding awkward or static postures? • How far will the object or person need to be moved?	• Using mechanical lifting aids whenever possible and making sure equipment is well maintained and safe to use • Only using equipment for its intended purpose • Planning sufficient rest periods so the muscles can recover • Avoiding situations where unsafe handling techniques may be used • Reorganising tasks, siting of equipment and placement of clients to reduce the need for awkward or static posture
I Individual	• Who will undertake the task? • What is their ability? • Are they suitably dressed to complete the task? • Have they received appropriate training? • Do they have Personal Protective Equipment *(PPE)* such as gloves if necessary? • How many people are required to complete the task safely?	• Individuals asking for help and guidance if they feel a manual handling task is beyond their capabilities • Only attempting moves that are within the physical capability of the individual *(especially if they are pregnant, have a health problem, injury or disability)* • Using more than one person if the load is too heavy or large. Good communication between the team is essential • Making sure the correct clothing is worn *(not too tight, with appropriate footwear)* • Making sure all staff have received proper training in the task they are being asked to complete
L Load	**If load is an object:** • How heavy is it? • What size and shape is it? • Where is its centre of gravity? • Could the load shift during the move?	• Reducing the object by breaking it down into smaller components • Making objects easier to hold by adding handles, insulating hot containers or wearing gloves • Not allowing the load to obstruct your view • Removing any unnecessary packaging
	If the load is a person: • Does the client have any level of independence? • What is the body shape, height and weight of the client? • Does the client have any behavioural issues? • Is the client limited by pain or injury? • Have the client's wishes been taken into consideration?	• Aiding people to move independently wherever possible • Reassuring the person being moved before you attempt to move them • Following the client's personal handling plan *(see page 19)*
E Environment	• Where will the task take place? • Is there enough space for the manoeuvre? • Are there any obstacles or hazards? • Can the layout, lighting or flooring be made safer?	• Checking the route is clear and any clutter is removed • Checking that the area which you are moving to is ready for the person or object • Thinking about where heavy or frequently used items are stored • Avoiding uneven or slippery floors • Making sure the working environment is well lit

Personal Handling Plan

In addition to the risk assessments that are completed for every manual handling activity, when handling people an individual personal handling plan should also be completed. A personal handling plan is a document that shows the individual risk assessment for a client and the recommended handling scheme. Having a personal handling plan is often a regulatory requirement and may be requested during inspections. It is very important that personal handling plans are followed by everyone who handles that individual client.

Personal handling plans benefit the client by providing:

- A consistent and well thought out approach to patient handling risk assessments
- A risk assessment that suits their individual needs
- A communication tool for all health care disciplines
- A written record of any changes in their care

Personal handling plans vary between organisations. The key information that should be included in any personal handling plan is:

Client's name Including any identification numbers

Client's height and weight *(this should be dated)*

Level of communication and comprehension

Handling constraints Including mental state, challenging behaviour, restricted normal movement, history of falls, tissue viability, pain and catheters

Assessment of specific skills Including walking, bathing, transfers, bed mobility and emergency evacuation

Level of assistance needed Including number of carers needed and use of equipment

Specific equipment needed E.g. Oxford Quickfit Medium Sling with shoulder straps on yellow and leg straps on red

When possible, consent should be gained from the client and should form part of their personal handling plan.

Personal handling plans should be reviewed regularly. They should also be reviewed if the client's abilities or circumstances change to make sure they remain suitable for the client.

On the spot risk assessments should always be completed by the carer before they move a client to check their personal handling plan is still suitable, e.g. if the client is unwell their usual personal handling plan may not be appropriate on that day. If any variation on the personal handling plan occurs this must be documented and kept with the plan.

The use of personal handling plans should also be monitored to make sure that carers are following them correctly.

Principles of Safer Handling – Moving Objects

Once you have established the manual handling task cannot be avoided and you have assessed the task and minimised the risks, you then need to follow the correct procedures for completing the task safely.

Testing if an Object is Safe to Lift

Before you even attempt to lift an object, you should always assess whether the load is safe to lift. You can do this by trying to slide the load from side to side with your fingers to see if it moves easily or not. If the load is too awkward in shape or you think it may be too heavy – GET HELP! You could also consider if the load could be broken down into smaller pieces to make it easier and safer to lift. If the load is sharp or excessively hot or cold, you will also require Personal Protective Equipment *(PPE)* such as gloves or an apron.

Lifting and Returning Objects to a Low Level

Using the following techniques when lifting and lowering objects will make each stage of the task safer.

Plan

- Stop and think about the task
- Use T.I.L.E. to assess and reduce any risks as much as possible

Position

- Get as close to the load as possible
- Position your feet shoulder width apart with the leading leg slightly forward
- If the load has an uneven centre of gravity, position yourself so that the heaviest side is closest to your body

Lift

- Keep your head up and your shoulders level
- Bend with your knees not your back
- Make sure the load is kept as close to your body as possible, ideally with your elbows at right angles
- Check you have a good grip using your whole hand and not just your fingertips
- Use your leg muscles to smoothly bring the load to waist height

Move

- Continue to keep the load close to your body
- Avoid twisting and sudden jerky movements
- Make sure you can see where you are going
- Keep carrying distance to a minimum

Lower

- Use the same principles used for lifting, with the movement reversed to smoothly lower the load
- Take care to avoid crush injuries to fingers and toes when putting the load down

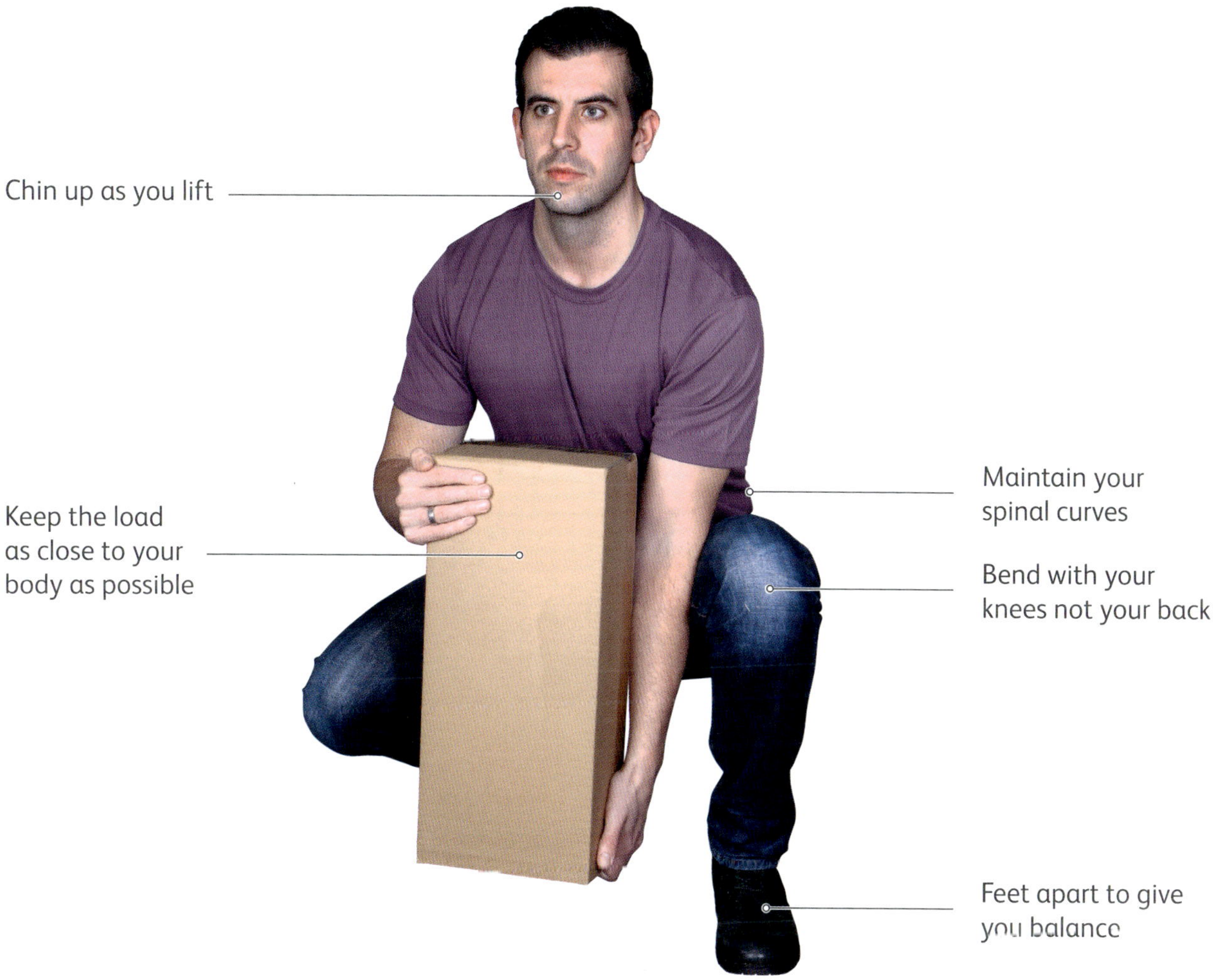

Feet Positioning

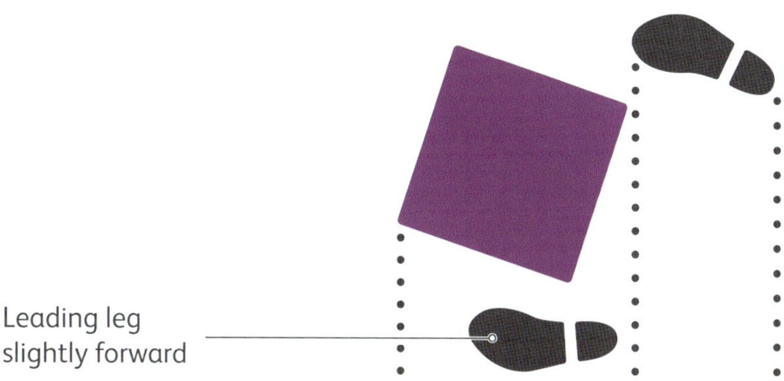

NOTE:
Always make sure you have a stable base when moving objects. If carrying more than one object, try to balance the load by carrying equal weight in each hand, e.g. when carrying 2 paint pots, carry one in each hand.

Things to Consider when Lifting and Moving Objects

Awkward and Asymmetrical Objects

When lifting and handling asymmetrical or awkward shaped objects, many of the same principles as handling regular shaped loads apply.

The main difference is the positioning of your hands on the load. When handling asymmetrical objects you should take hold of the opposite top and bottom parts of the load. If loads are too cumbersome for one person to handle, or you cannot see over the load, you should always seek help.

Working at Low Levels

If the height of the surface you are working on is at a low level and requires you to crouch down, it is far better to kneel than to bend and stoop to complete the task. If working at low levels for extended periods of time change your position regularly.

Lowering from a Height

Lifting from above your shoulder and reaching for loads away from your body puts a strain on your back and arms. Therefore if possible:

- Use a firmly placed stool or stepladder to get closer to the object
- Slide the object towards you and grip it securely with the whole hand and not just the fingertips. The item should be light enough to carry in one hand
- Reduce the weight of the object to be lifted by removing some of the contents if necessary
- Avoid lifting heavy objects from a height by storing them on a lower shelf in the first place

Pushing Loads

In most cases pushing a load is preferable to pulling a load. Before you begin the manoeuvre, make sure the equipment to be used is in correct working order, runs smoothly and is suitable for the task. The handles on the equipment should be between waist and shoulder height of the operator and close to their centre of gravity. Make sure there is enough friction between the floor and the operator's shoes.

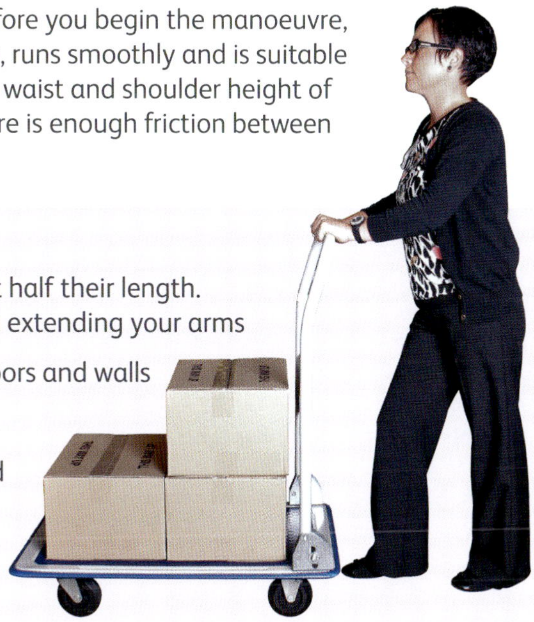

To push the load safely you need to:

- Bend your arms at the elbow and stretch them out to about half their length. This allows you to change the effort required by flexing and extending your arms
- Keep your hands where they will not get trapped against doors and walls
- Stand with one foot in front of the other
- Use your legs and the weight of your body to move the load
- Tighten your stomach muscles
- Bend your knees, maintain your spinal curves and lean into the push
- Keep your feet away from the load and maintain a walking pace

Pushing up a slope

Negotiating ramps and slopes may require help from another person.

If you need to change direction during a push manoeuvre – **DO NOT TWIST!** Always use your feet to change direction.

Pulling Loads

Before you begin the manoeuvre, ask yourself if the task could be done more safely by pushing rather than pulling the load. Ensure that the equipment being used is in correct working order, runs smoothly and is suitable for the task. The handles on the equipment should be between waist and shoulder height of the operator. Make sure there is enough friction between the floor and the operator's shoes.

To pull a load safely you need to:

- Keep your hands where they will not get trapped against doors and walls
- Put one leg behind the other to provide a secure base
- Maintain your spinal curves and keep your arms as close to your body's centre line as possible
- Be visually aware of your surroundings as you are not looking in the direction you are going
- Use your legs and the weight of your body to move the load
- Keep your feet away from the load and keep to a walking pace

If you need to change direction during a pull manoeuvre – **DO NOT TWIST!** Always use your feet to change direction.

Team Handling – Moving an Object

A load can be shared between individuals to reduce the weight. This means an object may be managed comfortably between a competent team, when it couldn't be managed by one person alone. However, lifting in a team does not mean that the weight of the object can be doubled by the addition of another person.

When possible, those lifting should be of the same height and build.

Make sure you have assessed and planned the move carefully. In particular make sure you know where you are taking the object, have checked the path is clear from obstruction, that there is space available to put the load down at the final destination and that the object and all team members will fit down the pathway chosen.

To work safely in a team you need to:

- Check there is enough space for the whole team to manoeuvre
- Check there are enough hand holds for each member of the team
- Nominate a member of the team to co-ordinate instructions
- Make sure there is clear communication between team members (consider effects of noise pollution when working in noisy environments – can everyone hear the instructions?)
- Be given clear instruction such as "Ready, steady, move" from the nominated person
- Bend your knees and straighten your legs in unison to keep the load balanced
- Make sure all members of the team look ahead to maintain their own spinal curves
- Position your body to avoid twisting
- Walk in small steps in unison – do not rush

If the load requires a member of the team to walk backwards, this person should not be the team leader. The team leader should always be facing forwards and should communicate clearly with the rest of the team so that nobody has to twist and look over their shoulder to see where they are going.

If you must turn whilst carrying the load – **DO NOT TWIST!** Under instruction from the nominated leader, turn in unison using your feet and not your torso.

Nominated person to give an agreed, clear instruction such as **"Ready, steady, move"**

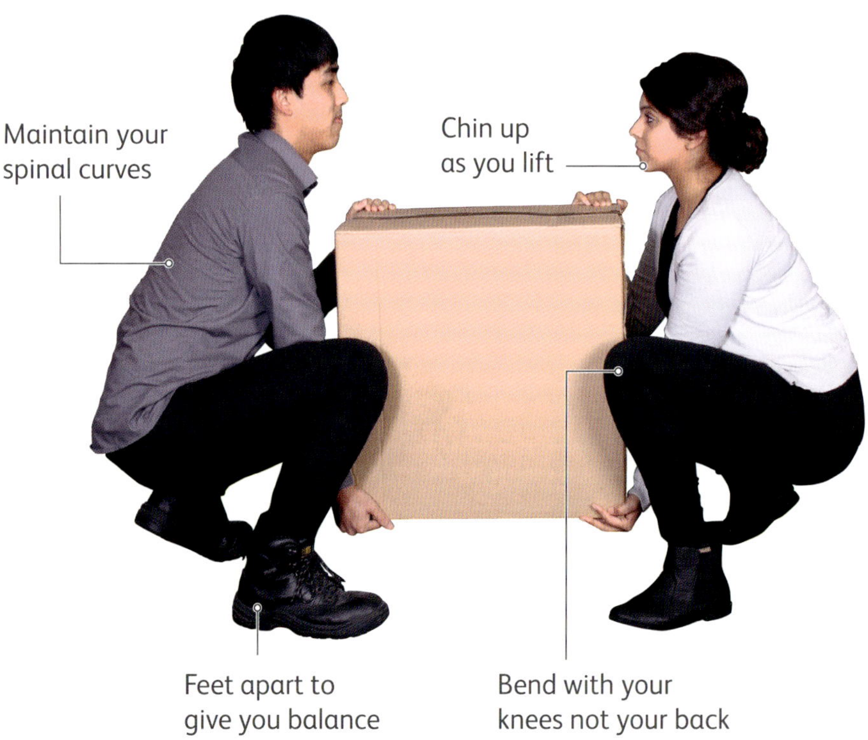

CONSIDER

One person could tilt the box to help the other person grasp the other side as part of the lift.

Equipment – Manual Handling of Objects

Types of Equipment

There is a wide range of equipment designed to be used for manual handling tasks. Getting the right equipment for the task and using it safely is vital. You should always check equipment visually before each use.

Low loader trolleys/platform trolleys

Advantages
- Takes larger loads
- Handles at waist height for pushing and pulling

Disadvantages
- Load shelf is low so items require low level lifting

Considerations
- Wheel size – larger wheels are easier to push
- Castors – improve manoeuvrability
- Tyres are more suitable for outdoors than solid wheels
- Is there a need for brakes? Can they be locked?
- A hydraulic platform would prevent stooping

Step ladders

Advantages
- Safer than overreaching for items
- Built for purpose

Disadvantages
- Involves working at height on a small surface area
- Can be unstable

Considerations
- Need to be used on flat even surfaces
- Handles on the side of the steps can help
- Steps with wider base of support will be more stable
- Straps are available to attach steps to surrounding structures
- Help stabilising the ladder may be required
- People who use ladders may require specific training in this area

High loader trolleys

Advantages
- Easy to load and unload without bending

Disadvantages
- Only suitable for smaller loads

Considerations
- Handles should be at waist height
- How secure is the load on the trolley?
- Wheel size – larger wheels are easier to push.
- Castors – improve manoeuvrability.
- Tyres are more suitable for outdoors than solid wheels. Is there a need for brakes? Can they be locked?

Small ladders/steps, e.g. elephant stool

Advantages
- Small and portable

Disadvantages
- Only raises the user up a small distance

Considerations
- Need to be used on flat, even surfaces
- Usually have a weight restriction of approximately 16 stone, so may not be suitable for all members of staff
- Should only be used if item being handled is light enough to hold in one hand

Conveyor belts/roller tracks

Advantages

- Reduces the need to carry loads long distances

Disadvantages

- Destination of the load is dictated by where the track is set up
- Risk of entanglement and entrapment in moving mechanical machinery

Considerations

- All operators need to be fully aware of the health and safety issues when using conveyor belts
- All hair should be tied back and clothing should not be hanging loose
- Noise pollution

Slide sheets

Advantages

- A friction free surface on which to push or pull items
- Can push and pull with less effort

Disadvantages

- May make the load unstable

Considerations

- Distance needed to push and pull – may only be suitable for short distances
- May restrict the weight of the load you can move

Mechanical hoists

Advantages

- Can lift heavier loads with minimum effort
- Decreases the number of staff needed to lift and move a load

Disadvantages

- Risk of entanglement and entrapment in moving mechanical machinery
- Hoists can be large which can restrict where they can be used

Considerations

- Specific training required
- Lifting equipment should be marked with safe working loads
- Is it suitable and stable for use?
- Needs to be positioned and installed to minimise risks *(from loads falling etc)*
- Requires servicing
- Noise pollution

Pallet trucks *(electric and manual)*

Advantages

- Less strenuous moving loads from one place to another
- Saves time and needs less manpower

Disadvantages

- Needs space to manoeuvre
- Takes up storage space when not in use

Considerations

- Specific training required
- Lifting equipment should be marked with safe working loads
- Is it suitable and stable for use?
- Requires servicing, lubrication and batteries
- Effects of vibration from the truck on the operator
- Noise pollution

Mechanical lifts, e.g. forklift trucks

Advantages

- Suitable for larger loads
- Can lift from a height

Disadvantages

- Requires lots of space to manoeuvre
- Many other health and safety issues to consider

Considerations

- Specific training required
- Effects of vibration from the truck on the operator
- Noise pollution

Door openers/magnetic door holds

Advantages

- Once opened they hold the door open for a length of time
- Eliminates the need to try and open a door whilst carrying a load

Considerations

- Holds on fire doors will need to release automatically when the fire alarm sounds
- Have doors that push open if regularly manual handling through doors, e.g. carrying dishes from a kitchen to the restaurant

Requirements for Testing and Servicing Manual Handling Equipment

Under the requirements of the Provision and Use of Work Equipment Regulations *(1998) – (PUWER)*, it is the employer's responsibility to make sure work equipment is well maintained, in efficient working order and is suitable for the task. When machinery in use has a maintenance log it must be kept up to date. If the equipment is used in conditions that may cause deterioration, the equipment should be inspected at suitable intervals.

Requirements for Testing and Servicing Lifting Equipment

Lifting equipment has its own set of regulations – Lifting Operations and Lifting Equipment Regulations *(1998) – (LOLER)*. An initial thorough examination of lifting equipment should be carried out before it is supplied. A further thorough examination should be completed at least annually by a 'competent person' who must then complete a written report. This inspection will be indicated by a label on the equipment. Anyone using it must check the date on the label and not use the equipment if it is out of date. The lifting equipment should be clearly marked with information on safe working loads *(for requirements for equipment used to lift people see page 34)*

Safe Storage of Manual Handling Equipment

When not in use, manual handling equipment should be stored safely to avoid the risk of obstruction and people tripping or falling over it. If possible, avoid storing manual handling equipment in conditions that could cause the equipment to deteriorate, e.g. storing outside where rain could rust trolley wheels.

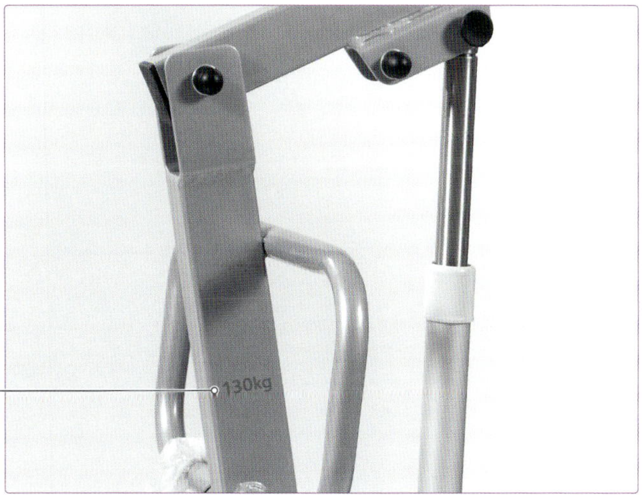

Safe working load

Personal Safety, Health and Dignity when Moving a Person

Competence

It is vital that you have the correct skills and knowledge to complete any manual handling task you carry out with a client. Attending manual handling training alone does not make you a completely competent manual handler. Competency improves with practice, review, feedback and reflection. Competency is not an indication of your pay grade, it is whether you have the necessary knowledge and ability to complete your job correctly.

'Competent'
Able, properly qualified, sufficiently trained, having the necessary skills and knowledge.

NEVER
Carry out a manual handling task that you do not feel competent to do. This could put you and your client at risk. Always seek advice and further training. Get advice and in-house training for any equipment you are not familiar with before using it.

Impact on a Person Moved Incorrectly

When moving a person it is essential you follow the correct procedures for the individual client. If you move a person incorrectly the results could be disastrous and could have a lasting impact on the client.

Incorrect handling of a person could cause them:

- Embarrassment and indignity
- Anxiety and worry
- Reluctance to move independently in the future
- Injury, or worsen existing injuries
- Paralysis
- Death

How Dignity and Health can be Preserved

By following the correct procedures and making sure handlers are suitably trained, the dignity and health of the client are far more likely to be maintained. Screens or curtains may need to be used if the move is in a public place. The correct use and maintenance of equipment is also crucial in making sure the client does not suffer any preventable humiliating or harmful accidents.

Clients who require manual handling may find it a frightening experience at first. **Communication** plays a vital part in making sure the client feels more relaxed about being moved.

For the client to participate in the manual handling task they must have an understanding of:

- What the task is and what they are expected to do
- What assistance the carers will provide
- Where the carers need to place their hands to assist
- What equipment will be used

When possible, carers should use a flat palm hand and place this over the client's shoulders, middle of the back or hips to lessen personal touch. Handling belts can also be used to provide a good handling grip whilst retaining more dignity for the client.

Use a flat palm hand on the middle of the back to lessen personal touch.

Consent

Manual handling does require personal touch and therefore you should be aware of the client's dignity and privacy at all times. Best practice would be to get the client's written consent as part of their personal handling plan *(see page 19)*. If written consent is not possible, carers should document verbal consent. Never presume consent, even if the client has experienced the activity before.

Infection Control

When moving a person the issue of infection control needs to be addressed to protect the carer and the client and prevent cross contamination. Hands are the most common way that bacteria are transferred from one place to another. Good hand hygiene is essential in reducing the risk of infections spreading.

Equipment for use with clients must either be washable, be able to be disinfected or should be single use.

Hand Cleaning Technique

Clean your hands before and after client contact and after contact with the client's surroundings.

This method can be used with soap and water or alcohol hand-rub. Make sure you have enough soap or hand-rub to cover all hand surfaces.

NOTE:
If your hands are dirty, contaminated with body fluids or you are caring for someone with diarrhoea or vomiting, you should always use soap and water.

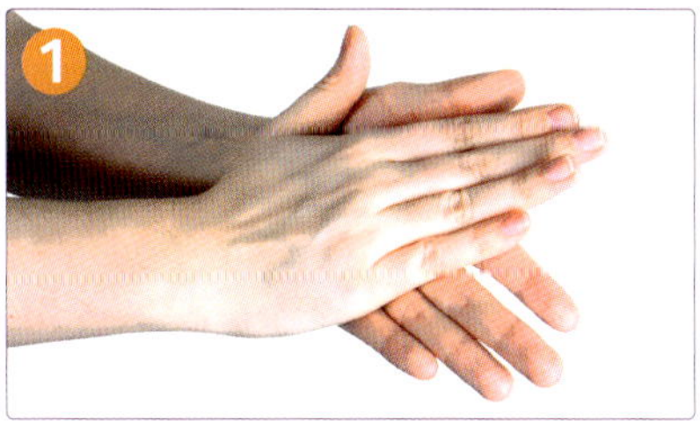

Rub palms together

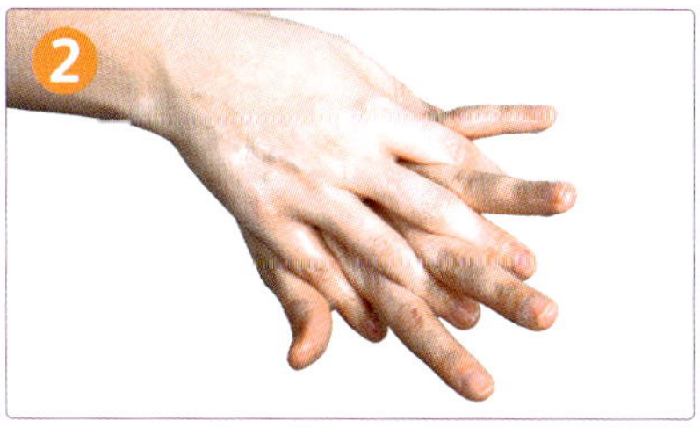

Rub each palm over the back of the other hand with fingers interlaced

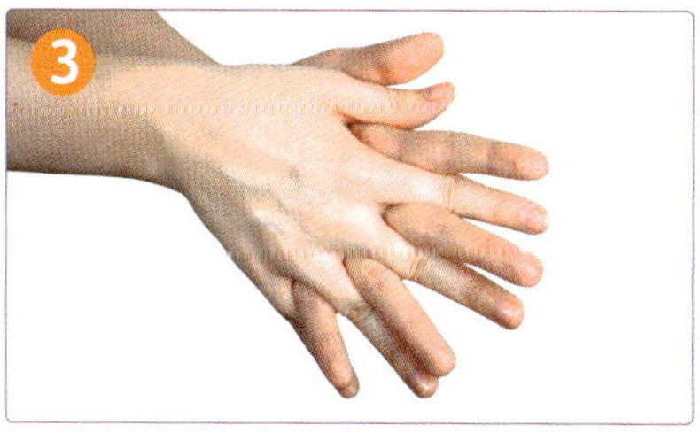

Rub palm to palm with fingers interlaced

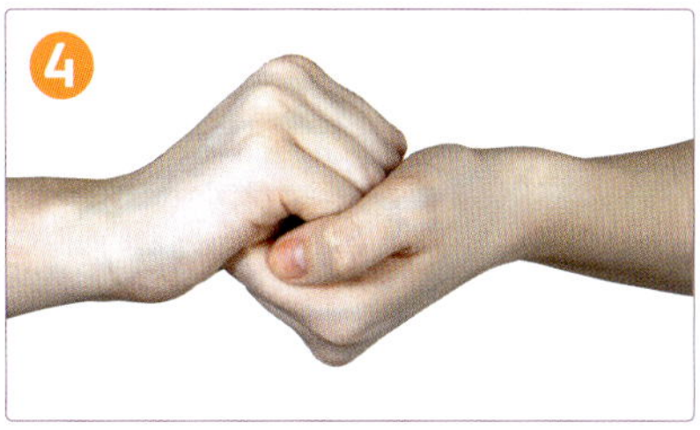

Interlock fingers and rub with backs of fingers to opposite palm

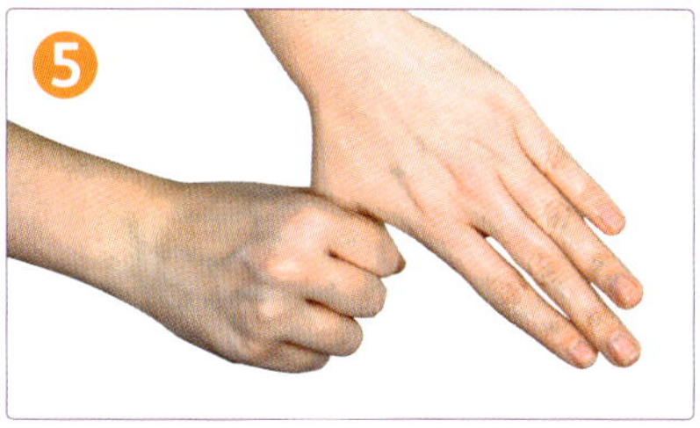

Rub around each thumb with the opposite hand

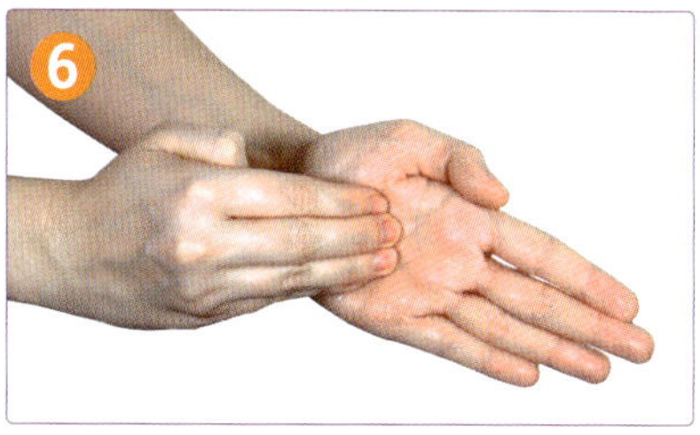

Rub in circles on each palm using fingertips

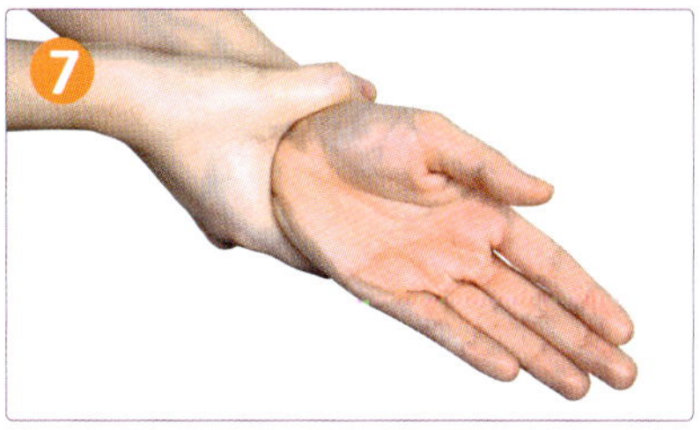

Rub each wrist with opposite hand

Turn any taps off with either your elbow or using a paper towel. Do not contaminate your hands by directly touching the tap.

When drying your hands after using soap and water it is preferable to use disposable paper towels. **NEVER** use a shared towel.

Functional Mobility

What is Functional Mobility?

Functional mobility is the ability to move from one location or position to another *(e.g. sitting to standing)* to participate in normal every day routines.

Levels of Mobility

- Independent – can move around purposefully with no assistance and is fully aware of their surroundings
- Partially Mobile – may need some assistance to complete a task
- Immobile – has little or no independent movement and requires complete assistance to move

Normal Movement

We have all developed basic techniques to carry out activities such as standing, getting up from sitting and getting out of bed. Although there will be small variations in our techniques they all share common features which are referred to as **normal movement**.

If we understand the normal movement for an activity, then we are able to employ the principles of safer handling to more effectively assist the client.

Normal Movement – Sit to Stand

To stand up from a chair you would normally:

- Move forward in the chair

- Bend forward at the waist
- Place your feet apart *(ideally one foot in front of the other)*

- Bring your head over your feet
- Start the upward movement to stand

Try standing without moving forward in the chair first. Compare how stable you feel standing with your feet close together and then positioned a shoulder width apart with one foot in front of the other.

To help the client keep as much independence as possible and reduce unnecessary handling, you should think about the normal movement of any activity and encourage the client to replicate this.

Things that might affect the client's normal movement:

- **Pain** – good pain relief is needed to help a client move to the best of their ability
- **Weakness** – client's weaknesses can be made worse by environmental factors such as low chairs or chairs without arms
- **Positioning of carer** – standing directly in front of the client will physically block their forward movement
- **Blocking knees** – knees shift forward during standing as part of normal movement so blocking the client's knees will prevent this
- **Blocking feet** – some clients will need to move their feet to regain balance so blocking their feet will prevent this
- **Balance** – client needs to be able to remain balanced

Verbal Prompts

Sometimes clients have the physical ability to perform an activity but have difficulty working out how to do it and the sequence of the task. In these cases giving verbal prompts may be all that is needed for the client to complete the task independently. Try verbal prompts first as this may eliminate the need to carry out any handling of the client.

Consider the normal movement of the activity and break it down into stages. Give verbal prompts and physical hints such as pointing and demonstrating to help the client.

Verbal Prompts for Sit to Stand

- "Place your hands on the arm rests"
- "Move closer to the edge of the seat"

- "Place your feet apart with one foot in front of the other"
- "Put your stronger leg slightly under the chair to help you push out of the chair"

- "Lean forward and put your nose over your toes"
- "Push with your hands to help you get up"

Levels of Assistance

A client's state of functional mobility can be determined by the level of assistance needed to complete an activity. If normal movement and verbal prompts have not helped the client to complete the task, they may need some further assistance. Different levels of assistance may be needed:

Minimal Assistance

- Involves some handling of the client as well as verbal prompts
- Client would be able to complete the majority of the activity
- Would usually require one carer
- Carer needs to get consent from client to assist them *(see page 29)*

Moderate Assistance

- Involves more handling assistance
- May require more than one carer
- Carer needs to get consent from client to assist them *(see page 29)*

Moderate assistance techniques are also used when the client is able to manage with minimal assistance most of the time but may be prone to, for example, losing their balance. If the nature of the client is unpredictable, moderate assistance techniques are used if it is assessed to be of less risk to the client and carer.

Two carers may also be needed to **AVOID DOING TWO TASKS AT THE SAME TIME**. For example when helping a client with toileting, one carer will be needed to support the client's balance whilst another carer assists with undressing for use of the toilet. One carer doing both tasks would be unsafe.

At no time during moderate assistance should carers be completely supporting the client's body weight. The client should have the physical and mental ability to assist with the activity. If this is not the case, then the use of manual handling equipment **MUST** be used to reduce the risk to carers and clients.

Equipment – Moving People Safely

Types of Equipment

There is a wide range of equipment designed to be used for moving people. Getting the right equipment for the task and using it safely is vital. Equipment for use with clients must be able to be washed or disinfected or should be single person use. Check the weight restriction of the equipment before use. **You should always check equipment visually before each use.**

Hoists

e.g. ceiling tracking hoist, mobile hoists, standing hoist

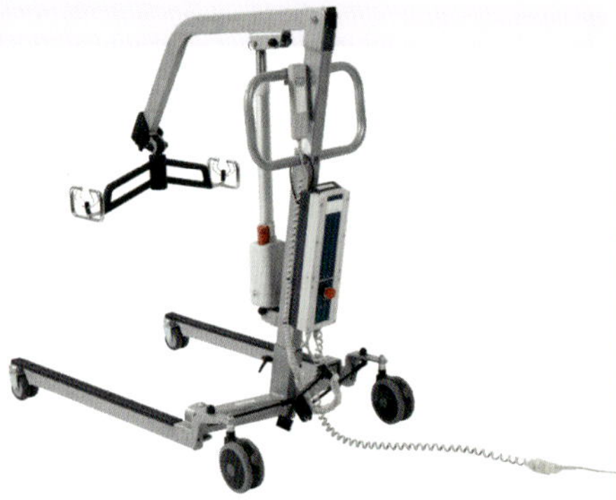

Advantages

- Suitable for moving people while minimising handling

Disadvantages

- Requires lots of space to manoeuvre

Considerations

- Specific training required

Slide Sheets

Advantages

- Provide a friction free surface on which to push and pull so less effort needed

Disadvantages

- May make the person being moved unstable

Considerations

- Distance needed to pull or push

Straps

To be attached to slide sheets

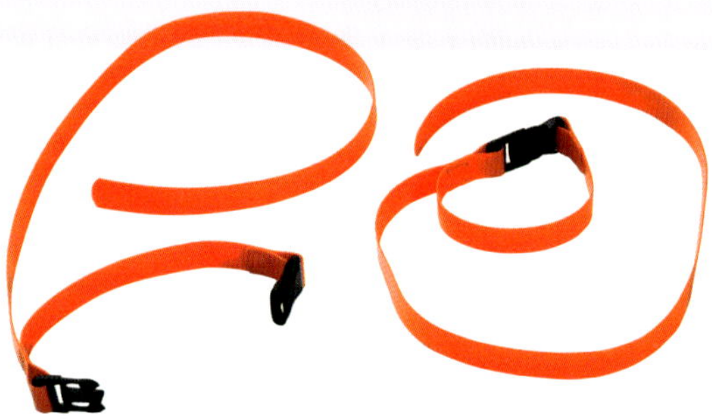

Advantages

- May reduce overreaching

Disadvantages

- May keep carer away from client

Considerations

- Could be used to bring a client closer before performing further handling techniques

Handling Belts

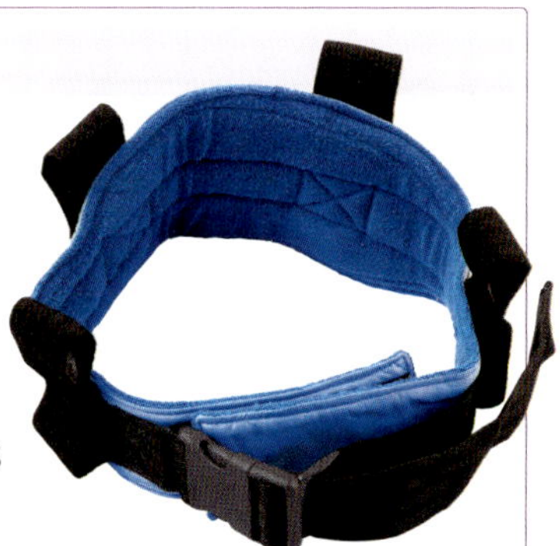

Advantages

- Provide a hand hold for the carer to prevent them holding the client's clothing or applying pressure to skin surfaces
- Can be used to assist walking or transfers

Disadvantages

- Need to be fitted firmly to be effective which might be uncomfortable for clients

Considerations

- Come in a variety of sizes and padding. Must **NOT** be used as a lifting aid

Transfer Boards

Advantages
- Useful as an aid in seated transfers
- Shape designed to accommodate arms on chairs
- Helps the client to transfer independently from one seat to another without the need to stand

Disadvantages
- Chairs must be able to be positioned close together
- Client requires good upper body strength and good balance in sitting

Considerations
- Usually have a weight restriction of about 20 stone, so may not be suitable for all clients

Stand and Turn Transfer Aids

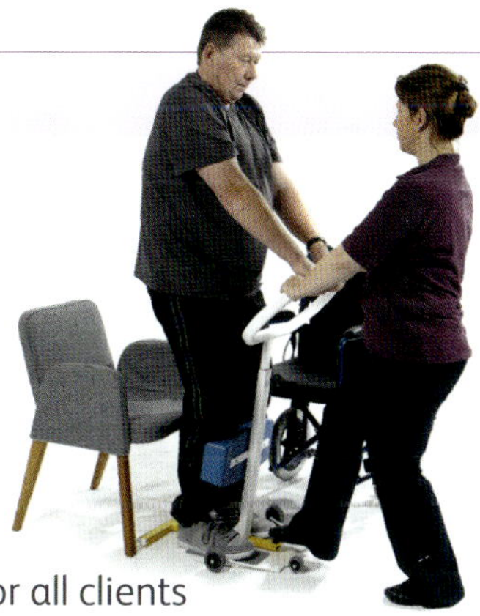

Advantages
- Suitable for moving people while minimising handling

Disadvantages
- Most only turn on the spot to help transfers
- Requires client to be able to weight-bear and have upper body strength

Considerations
- Usually have a weight restriction of about 16 stone, so may not be suitable for all clients

Turntables

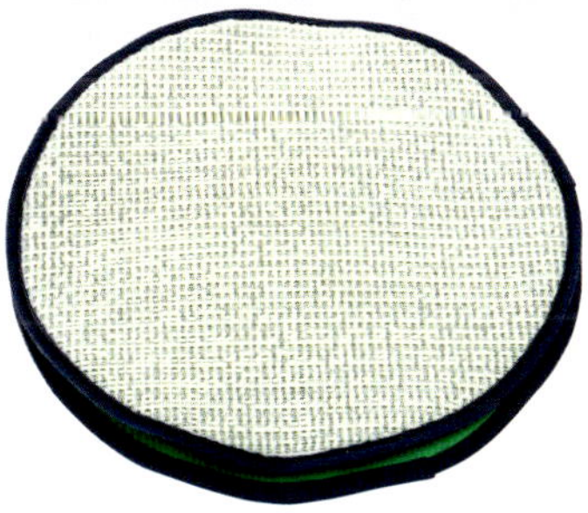

Advantages
- Useful as an aid to transfers in either sitting or standing

Disadvantages
- Most only turn on the spot to help transfers
- Requires client to be able to weight-bear and have upper body strength

Considerations
- Usually have a weight restriction of about 16 stone, so may not be suitable for all clients

Profiling Beds

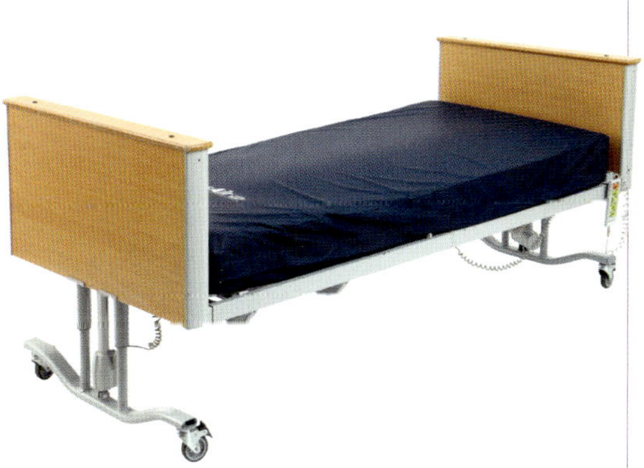

Advantages
- Bed height and seating position can be adjusted electronically. This improves carer's posture when care giving and may also help the client to be more independent in their bed mobility and avoid unnecessary handling

Disadvantages
- Expensive

Considerations
- Usually have a weight restriction of about 24 stone but versions are available for plus sized clients

Emergency Lifting Cushion

Advantages
- Able to lift the client from the floor in seated or lying positions without carers needing to lift them
- Some brands can lift up to 70 stone
- Portable as battery powered so can be used indoor and outdoors

Disadvantages
- Must be kept on charge so ready when needed

Considerations
- Used by ambulance services and some companies who provide support for clients who have fallen
- Costly but a valuable investment if the client has regular falls

Wheelchairs

Advantages
- Helps a client with limited exercise tolerance or balance to be moved safely over longer distances

Disadvantages
- Requires space to manoeuvre

Considerations
- Usually have a weight restriction of about 24 stone but versions are available for plus sized clients.
- Some specific training needed

Requirements for Testing, Servicing and Examining People Moving Equipment

Provision and Use of Work Equipment Regulations *(1998) – (PUWER)*

Equipment used at work should be:

- Suitable for purpose
- Regularly maintained and documented in maintenance logs
- Regularly inspected
- Only used by people with adequate training

Lifting Operations and Lifting Equipment Regulations *(1998) – (LOLER)*

Equipment used to lift a person should be:

- Inspected thoroughly every 6 months by a competent person and have an in date inspection label
- Suitable for purpose
- Only used by people with appropriate training
- Marked with information on safe working loads

NOTE:
The weight of a person should form part of the client's personal handling plan *(see page 19)*. Lifting equipment should not be used to move a person if they weigh more than the safe working load for that piece of equipment.

Principles of Safer Handling – Moving a Person

Unsafe Practice

As a result of injuries to clients and carers and due to a better understanding of our body mechanics, some manoeuvres have been nationally recognised as **UNSAFE** and should **NOT** be used.

Unsafe manoeuvres – issues for the client:

- Prevent client from participating in the activity and stop their normal movement
- Can involve pulling on the client's shoulder region so have a high risk of trauma

Unsafe manoeuvres – increased risk of injury to the carer due to:

- Carer's posture being compromised as a result of twisting and stooping
- Carer taking the majority of the client's body weight

CAUTION
If a client has fallen they should not be lifted manually from the floor. Use a hoist.

The following techniques have been nationally recognised as UNSAFE and should NOT be used:

Drag lift

Carer places a hand or arm under the client's armpit and pulls or drags them to move them.

Bear hug

Carer stands in front of client and client puts their arms around the carer to 'hug' them. Carer then assists the client to transfer to a different position.

Australian lift

A type of shoulder lift used to move patients. A carer stands either side of the client and places their shoulder under the client's armpit. The two carers hold each other's hands under the patient's thighs creating a seat for the client. Client is then lifted in this seated position and moved to a different position.

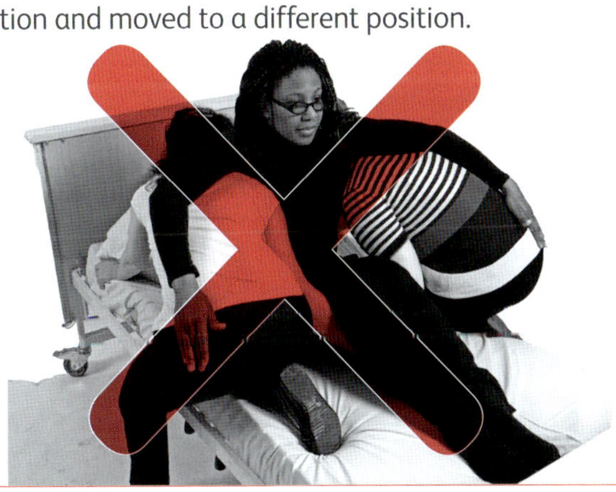

Top and tail

Two carers lift the full weight of the client. One carer stands behind the client and puts their arms under the client's armpits. The other carer takes the weight of the client's legs.

Principles of Safer Handling Techniques

Power Position

The stability of the carer's body when moving a client plays an important role in protecting the carer's and the client's health and safety. Carers should make sure that they have a wide base of support with their feet apart to give them better balance. Carers should bend their knees, keep their chin up and try to maintain the normal curvature of their spine. To decrease the pressure on their spine, carers must remain close to the client when helping them to move.

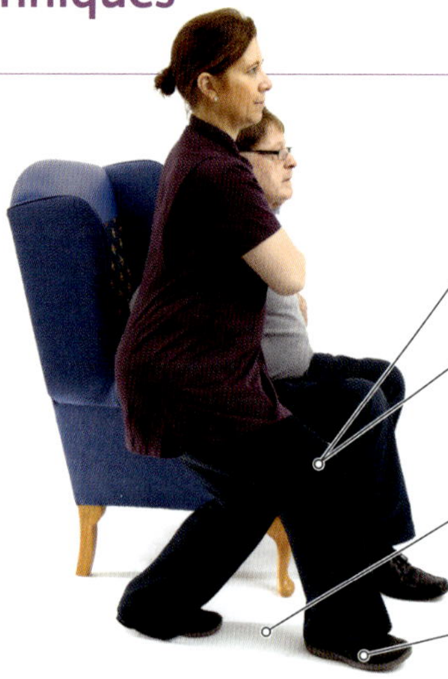

- Relaxed knees
- Bend with the knees and not the back to get close to the load
- Feet apart with a wide base of support
- One foot ahead in the direction of movement

There are a number of activities that carers will routinely use when helping clients to move. The following techniques are examples that show how the principles of safer handling can be used. It may not always be possible to reproduce these techniques exactly due to the environment, for example if there are bedside cabinets or limited access to the bed, or due to the individual's needs. Use these techniques as a guide and if you are required to make alterations, use the principles of safer handling to ensure your safety.

The following techniques presume that:

- Normal movement and verbal prompts have already been attempted
- Risk assessments have been completed
- Personal handling plans have been consulted
- Consent has been gained from the client
- Training has been given to the carers
- Infection prevention and control has been addressed
- Equipment has been checked, is suitable for the individual and is in working order
- Adequate space is available to complete the full manoeuvre and the route is clear

This symbol indicates the number of carers required.

Sit to Stand

Minimal Assistance

- Put flat palm hand across client's back to encourage them to move forward

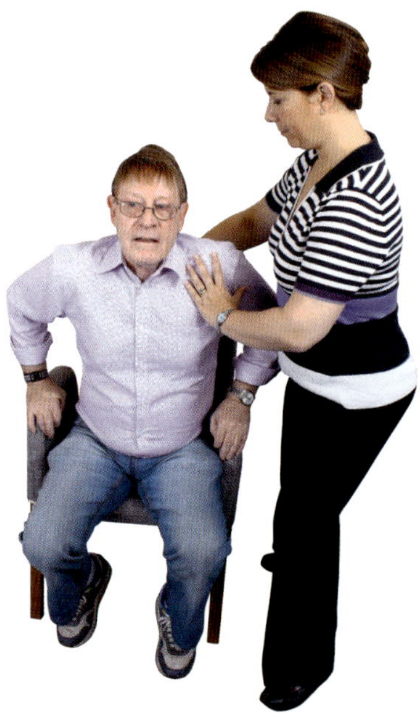

- Encourage client to use hands on armrests to participate
- Use verbal prompts throughout

Moderate Assistance Facing Same Way as Client

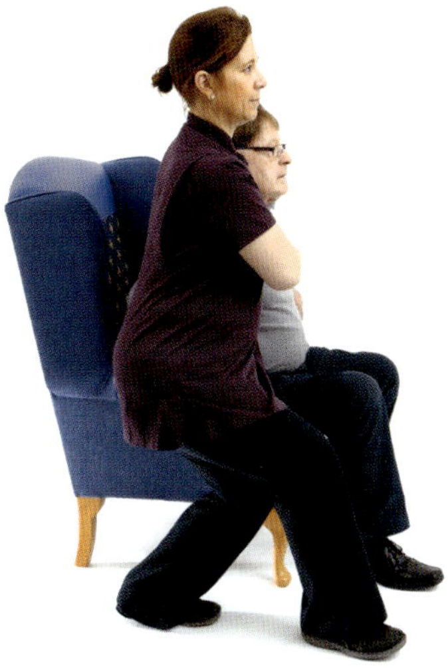

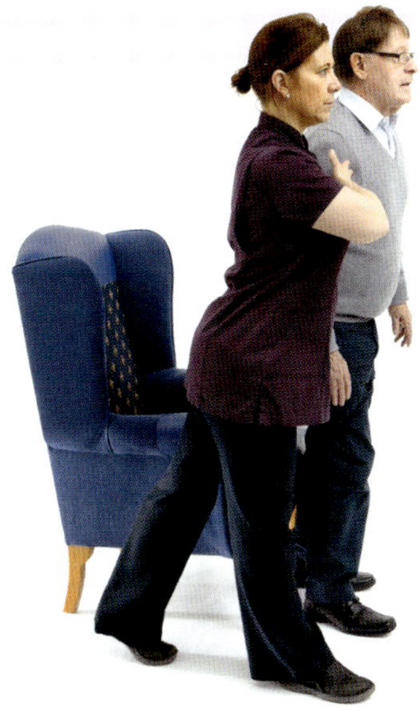

- Put flat palm on client's back to encourage them to move forward
- Adopt power position
  - feet apart with a wide base of support
  - one foot ahead of the other in direction of travel
  - relaxed knees
  - bend with the knees not the back
  - keep close to the client
- Face forward and avoid twisting

- Place foot level with client's front foot
- Bend knees sufficiently so your centre of gravity is level with client's centre of gravity
- Place hand on front of client's shoulder to encourage upward movement and provide support

- Encourage client to use hands on armrests to participate
- Use command such as "Ready, steady, stand" *(slight rocking motion can be used on "Ready, steady" to increase momentum)*
- Transfer your weight from rear foot to front foot as client stands

Moderate Assistance Facing Opposite Directions

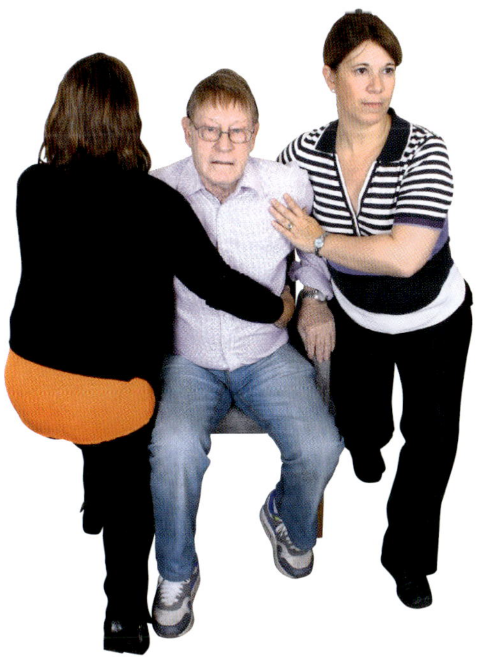

This move prevents crossing of carers' arms.

- One carer should be positioned on either side of the client
- One carer should face forwards with their arm across the client's back. One carer should face backwards with their arm across the client's front
- Adopt power position
  - feet apart with a wide base of support
  - one foot ahead of the other
  - relaxed knees
  - bend with the knees not the back
  - keep close to the client
- Avoid twisting
- Bend knees sufficiently so your centre of gravity is level with client's centre of gravity
- Encourage client to use their hands to push on the armrests to stand
- Transfer weight from foot nearest back of the chair to foot by the front of the chair as client stands

Using a Handling Belt with Moderate Assistance

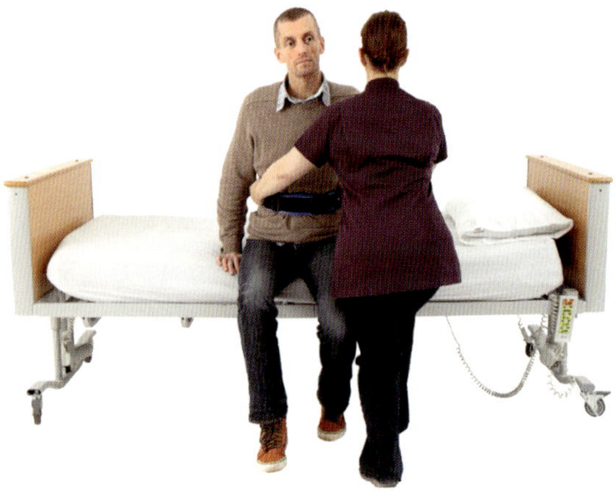

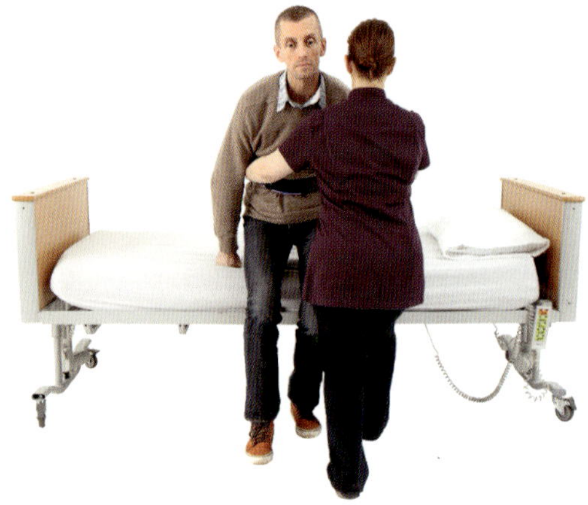

- Secure belt around client's waist using manufacturer's instructions
- Adopt power position
  - feet apart with one foot slightly under bed
  - one foot ahead of the other
  - relaxed knees
  - bend with the knees not the back
  - keep close to the client
- Place arm across client's chest and hold handling belt to encourage forward movement of client
- Place other hand on back of client's shoulder to encourage their forward movement
- Face backwards to avoid twisting your body

- Bend knees sufficiently so your centre of gravity is level with client's centre of gravity
- Encourage client to use their hands to push on the bed to stand
- Transfer your weight from front foot to rear foot as client stands

CONSIDER

Could the bed height be adjusted to make movement easier?

Using a Handling Belt with Moderate Assistance

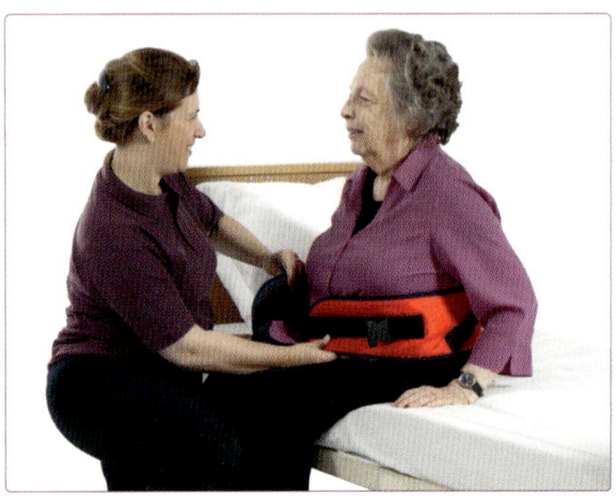

- Secure belt around client's waist using manufacturer's instructions

- Both carers face the client
- Place hand on the front of client's shoulder joint to encourage upward movement of client
- Adopt power position
  - feet apart with one foot slightly under bed
  - relaxed knees
  - bend with the knees not the back
  - keep close to the client

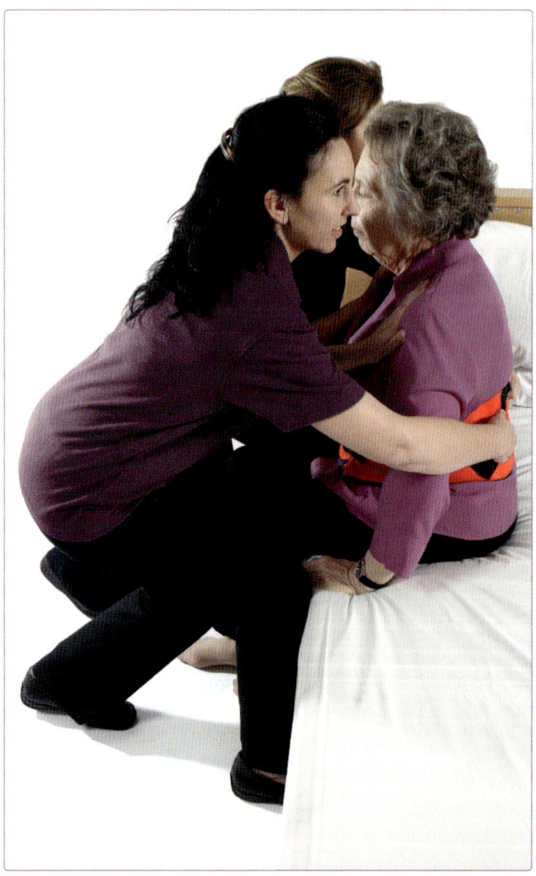

- Keep head up
- Avoid twisting
- Put hand behind client's back and hold nearest handle on the belt **CARERS MUST NOT CROSS HANDS**
- Bend knees sufficiently so your centre of gravity is level with client's centre of gravity
- Encourage client to use their hands to push on the bed to stand
- As client lifts their head to stand put a forward pressure on the handle to assist the stand. Do **NOT** lift the client
- Transfer weight from front foot to rear foot as client stands

Assisted Shuffle

Assisted Shuffle Forward in Chair

- Ask client to lean forward in seat and hold armrests
- Kneel on one knee in front of client, slightly to one side
- Place one hand on client's hip and other hand on client's opposite knee
- As client rocks to one side, put slight pressure on client's knee and ease the raised hip forward
- Repeat as needed

Stand to Sit

Minimal Assistance Facing Towards Client

- Adopt power position
  - feet apart with a wide base of support
  - one foot ahead of the other in direction of travel
  - relaxed knees
  - bend with the knees not the back
  - keep close to the client
- Put flat palm hand on front of client's pelvis and back of their shoulders to encourage them to bend forwards
- Encourage client to reach for chair arms

- Bend knees and transfer weight from rear foot to front foot as client lowers into the chair

Assisted Walking

Moderate Assistance Without a Walking Aid

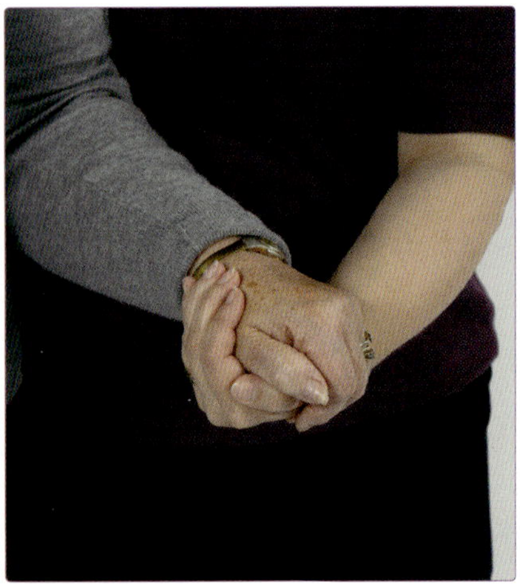

- Hold client's nearest hand at about hip height
- Avoid interlocking thumbs with client as this can cause injury to the thumb joint
- Face direction of travel

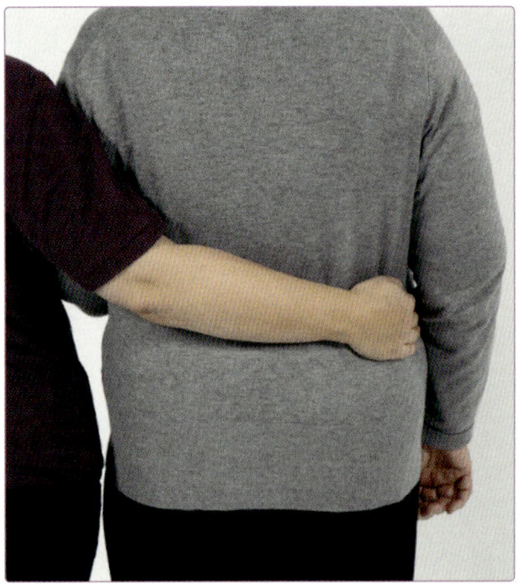

- Place flat palm hand across client's pelvis or shoulders
- Keep close to client
- Allow space for you and client to walk without stepping on each other
- Walk at a suitable speed for the client

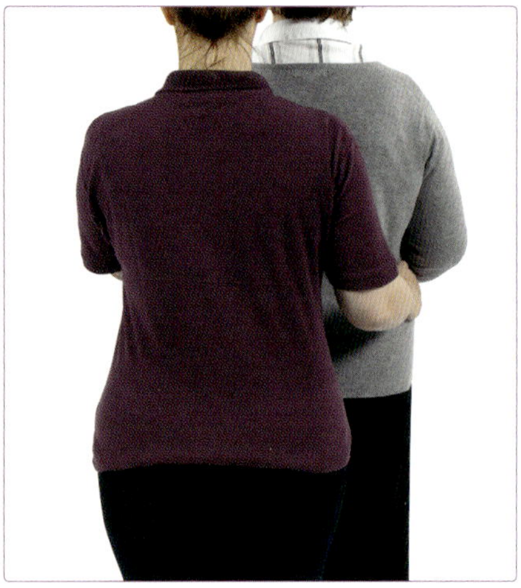

- If walking through a narrow space move behind client and hold their pelvis with both hands

CONSIDER

This technique can be adapted for use with a handling belt or if the client has a walking aid.

Transfer

From Chair to Wheelchair Using a Transfer Board

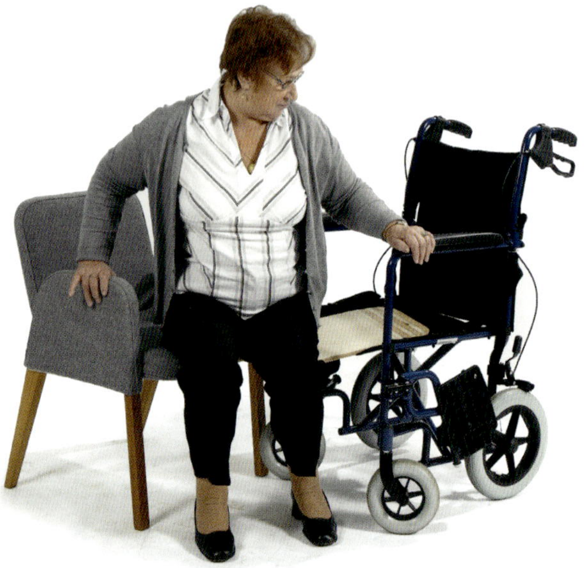

- Follow manufacturer's instructions for transfer board
- Maintain carer's spinal curves as they insert the transfer board
- Position the wheelchair *(wheelchair can be positioned more to the side of the chair if there are no arms on the chair)*
- Client reaches across to wheelchair arm

- Client shuffles across the board to the wheelchair

CONSIDER

Are the wheelchair brakes on?
Are the footrests in the correct position?

Using a Stand and Turn Transfer Aid

There are a variety of stand and turn transfer aids available. They are designed to assist carers in transferring clients from one seated position to another.

Follow manufacturer's instructions for the specific stand and turn transfer aid. The following points are meant only as a general guide:

1

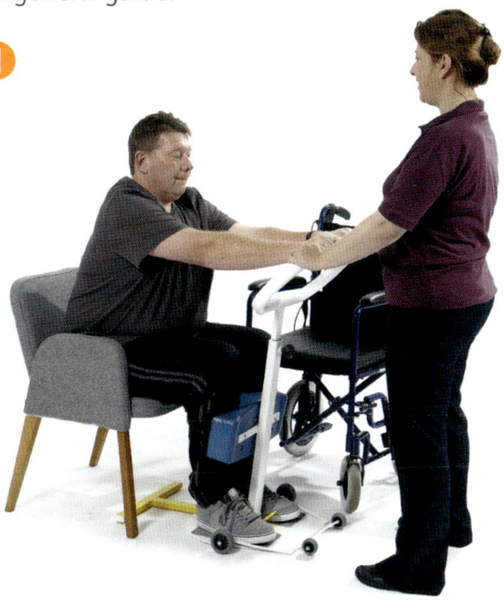

- Client must be able to pull themselves up into a standing position and balance without support
- Put stand and turn transfer aid in front of client
- Adjust height of handlebars and shin pads
- Client should sit on front of chair with feet on stand and turn transfer aid foot plate
- Wheel the stand and turn transfer aid so that the client's shins are just in front of the shin pad. The shins should not be pushed against the pads
- Apply the brake

2

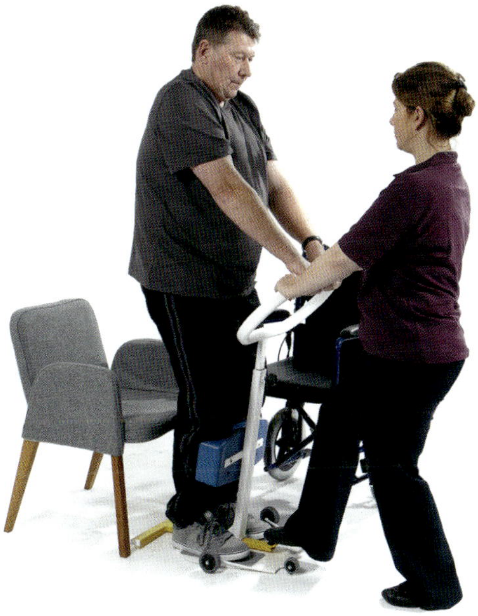

- Encourage client to pull themselves into the standing position using the handlebars to help them

(A second carer could be used to help the client if required. A handling belt may be used)

3

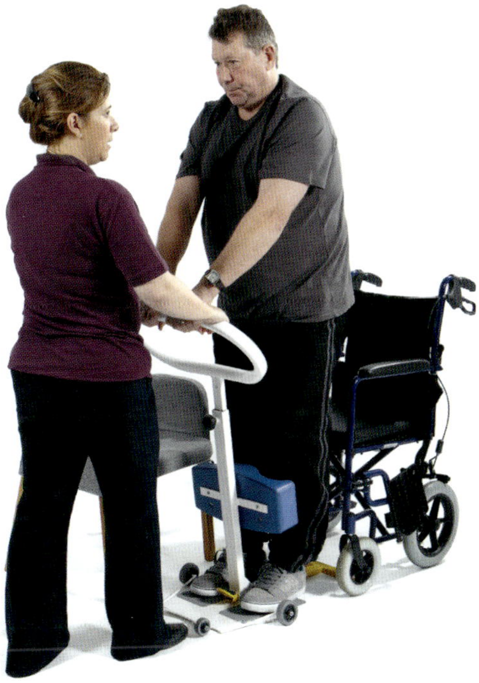

- Release the brake and rotate the stand and turn transfer aid

4

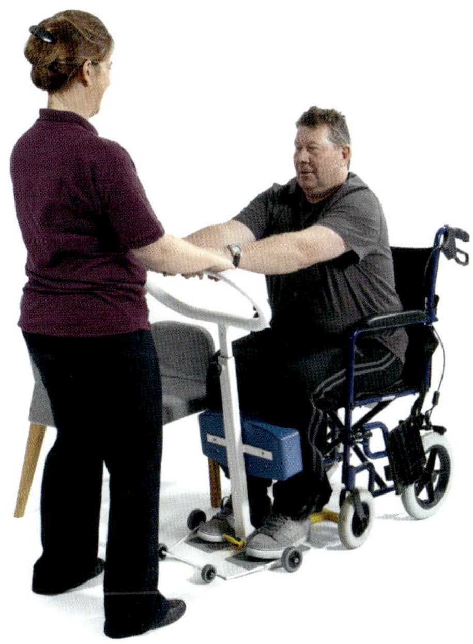

- Encourage client to sit using the reverse of the above technique

Bed Mobility

Assisting Into Bed

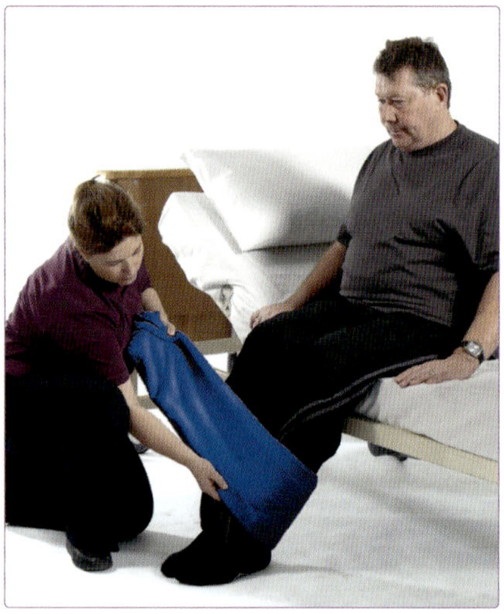

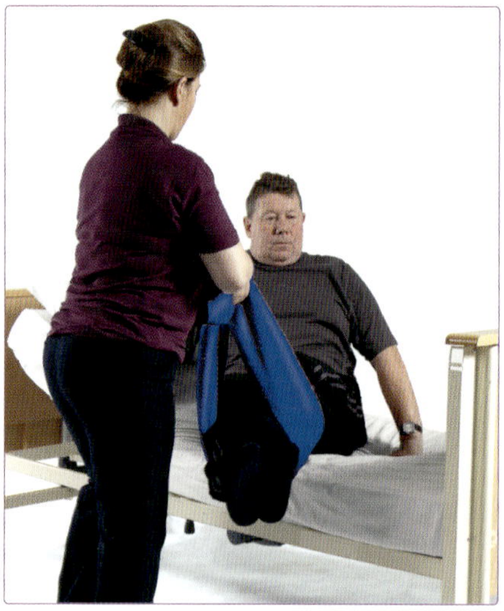

- Client should sit on edge of the bed
- Kneel to the side of the client
- Support the client's ankles using a handling sling
- Keep close to the client

- Using handling sling assist client to raise their legs
- Encourage client to rotate their body using their arms
- Two carers may be required. One to support the legs and one to support the top of the body

CONSIDER

Could the bed height be adjusted to make movement easier?

A slide sheet could be added to this technique to reduce friction. It could also be used to slide the client's legs to the middle of the bed to avoid lifting the load away from your body.

Assisting Out of Bed

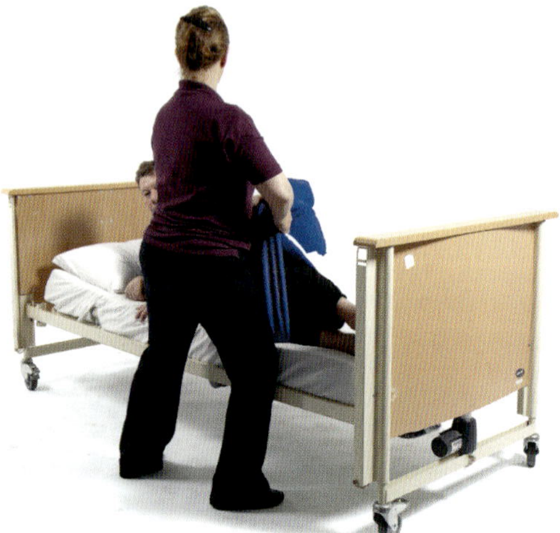

- Place feet apart
- Relax knees
- Maintain your spinal curves
- Client supports themselves on their elbow
- Support the client's legs using a handling sling
- Keep close to the client

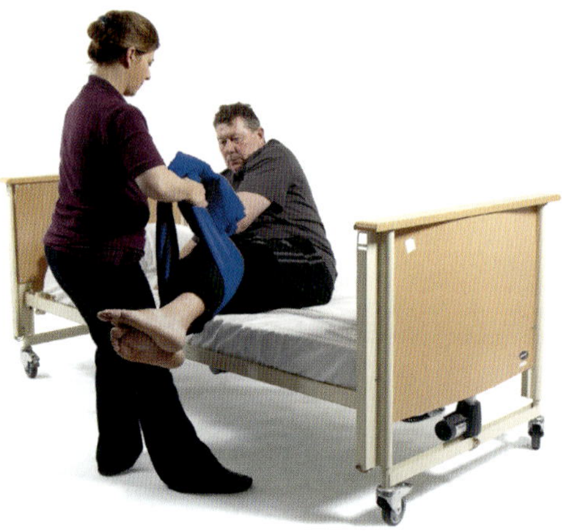

- Assist client to raise their legs with handling sling
- Encourage client to rotate their body
- Transfer your weight from one side to the other as you help the client's legs to move out of bed
- Two carers may be required. One to support the legs and one to support the top of the body

CONSIDER

Could the bed height be adjusted to make movement easier?

A slide sheet could be added to this technique to reduce friction.

Insertion of a Slide Sheet with Client in Lying Position

Slide sheets provide a reduced friction surface and need to work in two layers to be effective. There is evidence that slide sheets can also reduce the shearing forces on client's skin and therefore lessen the risk of skin breakdown.

Slide sheets can be inserted by rolling the client, but this can result in unnecessary handling. Slide sheets can be inserted over or under the bed sheet using the same technique.

1

- Roll the slide sheet from each edge to the middle

2

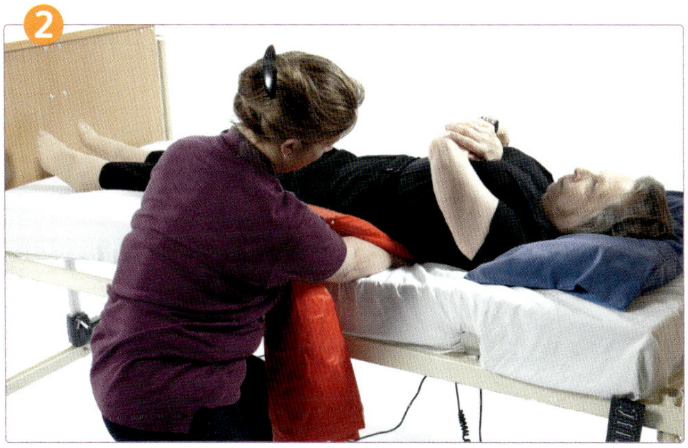

- Keep your hand under the slide sheet so your hand doesn't come into direct contact with the client
- Insert slide sheet under client's natural body curves

3

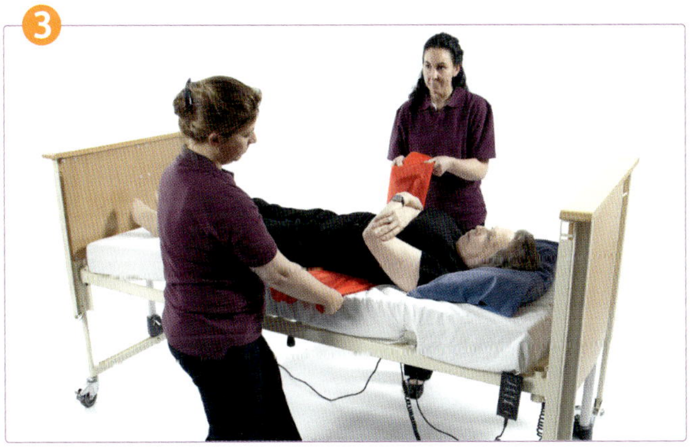

- Carer on the other side of the bed should pull slide sheet through so slide sheet is on each side for carers to hold

4

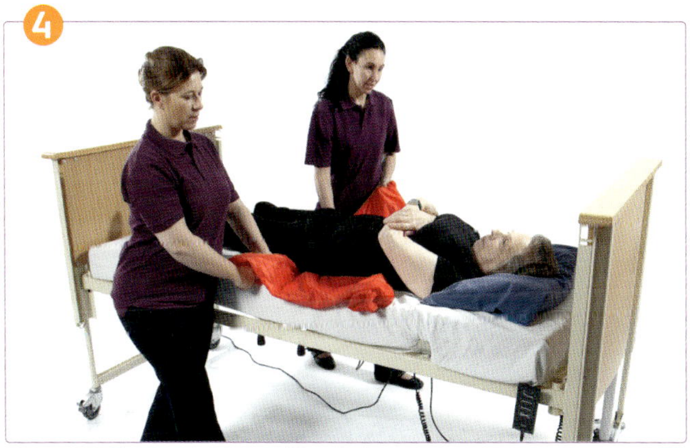

- Each carer should unroll half of the slide sheet. Use a light push/pull motion between the carers

5

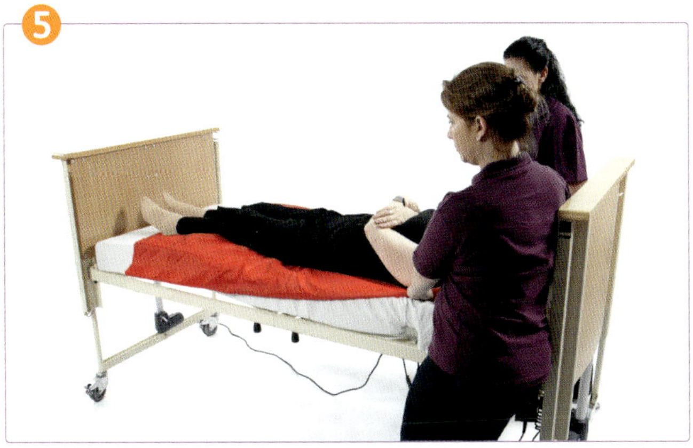

- Repeat the process to unroll the other half of the slide sheet
- Do not worry about occasional wrinkles in the slide sheet as it will still work effectively and will not be under the client long enough to cause skin damage

CONSIDER
Could the bed height be adjusted to make movement easier?

Turn with Slide Sheet

This technique can be used when the client is unable to assist and is more dependent. The slide sheet helps with this activity when used with two carers. The slide sheet can be placed directly under the client or under the bed sheet. This technique also avoids the client being rolled onto the edge of the bed as the slide sheet turns the client on an axis. This prevents further handling activity to bring the client back to the centre of the bed.

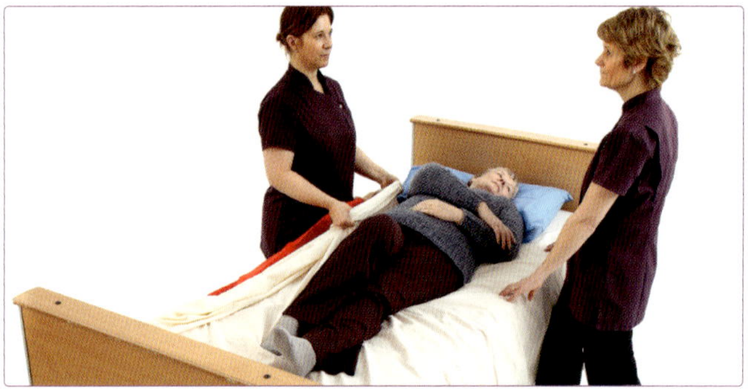

- Insert slide sheet under the bed sheet *(using a similar technique to that on page 43)*
- One carer should stand on either side of the bed with one foot in front of the other
- One carer should hold the bed sheet near client's pelvis area
- Encourage client to turn their head in the direction of the roll and raise their opposite knee

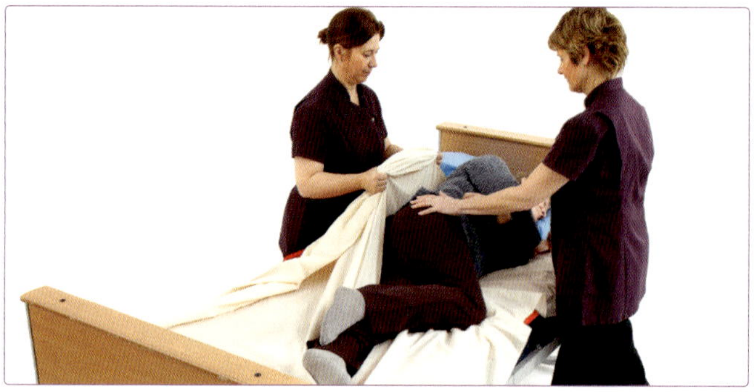

- One carer should pull the bed sheet towards them and upwards
- Other carer should make sure the client does not roll off the bed

CONSIDER

Could the bed height be adjusted to make movement easier?

Check there are no signs of wear on the sheet before use.

Slide up the Bed Using a Slide Sheet

The ideal position when sliding a client up the bed is to stand at the head of the bed.

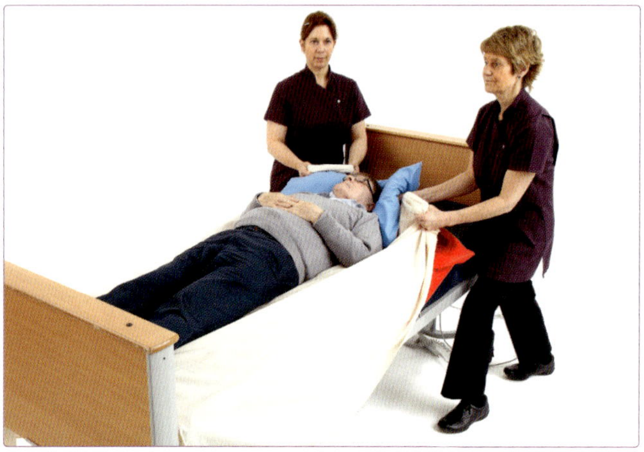

- Insert slide sheet under the bed sheet *(using a similar technique to that on page 43)*
- One carer should stand at the top corner on either side of the bed
- Both carers should angle their body diagonally. Hold bed sheet at the head of the bed
- Adopt power position
  - feet apart with a wide base of support
  - one foot ahead of the other
  - relaxed knees
  - bend with the knees not the back

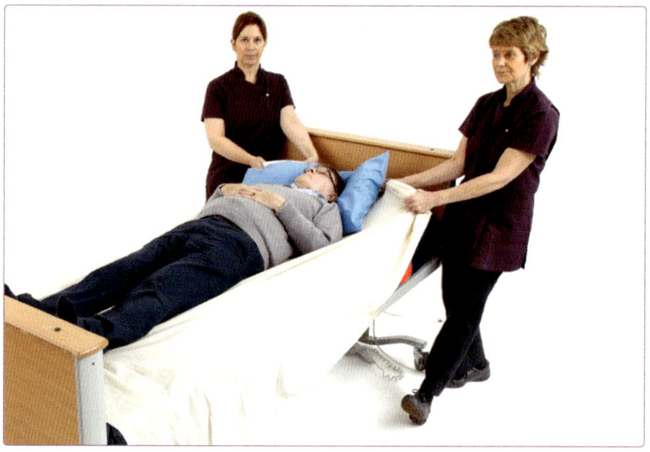

- Use command such as "Ready, steady, slide"
- Transfer your weight from front leg to back leg as you slide the sheet up the bed
- **Make sure you are not tempted to lift the client!**

CONSIDER

Could the bed height be adjusted to make movement easier?

Check there are no signs of wear on the sheet before use.

Slings and Hoists

Slings

There are a wide variety of slings in different sizes and materials to suit all users.

Every sling should have a label which indicates:

- Size
- Weight restriction
- Washing instructions
- Serial number to help track the sling

If the sling's label is not readable then the sling has come to the end of its working life and should be replaced. The carer has a responsibility to check the sling before each use, e.g. to check if the stitching is coming loose. Special slings are available for a variety of functions such as showering and some are designed to be left under the client whilst seated.

Make sure straps on slings are even under the client after fitting. Perform a tug test to make sure there are no folds under the client which might cause them to shift balance as hoisting commences.

Hoists

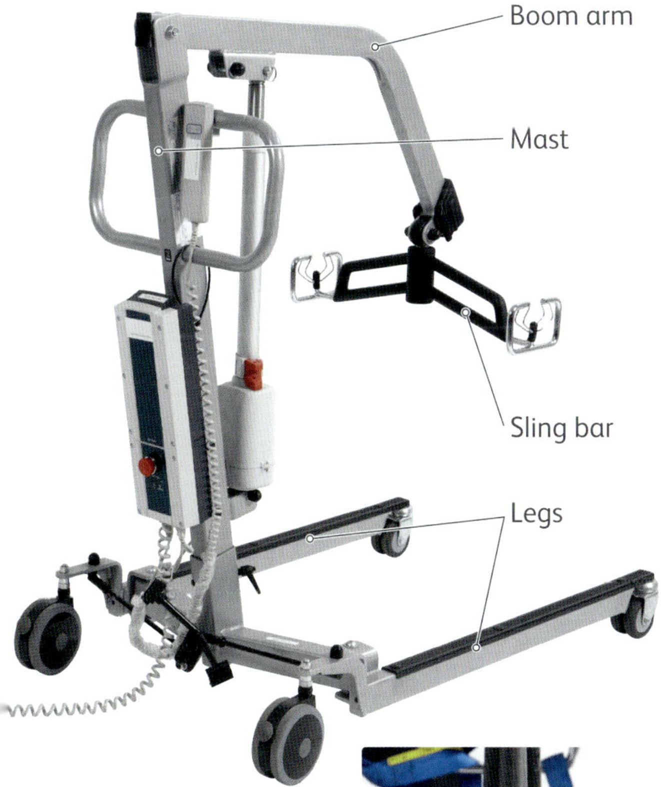

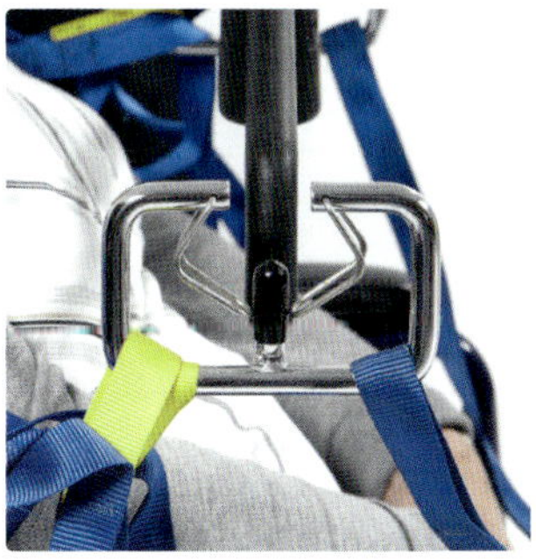

Sling attachments vary. Loop fixtures or secure button attachments are commonly used.

There are a wide variety of hoists available.

Most hoists have some common features including:

- Emergency stop button
- Emergency lowering system
- Adjustable width legs
- Ability to hoist from the floor

All hoists should clearly state the weight restriction of the equipment and have an in date service label. Anyone using it must check the label and not use the equipment if it is out of date. Mobile hoists for plus sized clients are available up to 300kg. For weights above this, ceiling or gantry hoist systems are recommended. For requirements for testing and servicing hoists see page 34 PUWER AND LOLER.

It is important to make sure you have the correct style sling for the hoist being used and that the equipment and configuration matches the client's personal handling plan (see page 19).

It is good practice to use 2 carers for hoisting but depending on the circumstances more carers may be needed, e.g. hoisting from the floor following a fall. Hoists are available for single-handed care but must be risk assessed as appropriate for the client.

It is important that **all** hoisting activities are risk assessed and that carers receive specific training before using individual hoists.

Good communication between the carers and the client is essential. Hoisting can be a frightening experience for the client and clear communication can reduce their concerns.

Hoisting

Fitting a Sling to a Client in a Wheelchair

1

- Assist or instruct client to sit forward in wheelchair
- Place main section of sling down client's back
- Keep centre of sling to centre of client's back and well down the base of their spine

2

- One carer should kneel in front of the client
- Carer on floor should place client's foot on their knee to raise client's thigh from the chair. This keeps the carer's hands free

3

- Assist or instruct client to lean to the side to increase access to put sling under client's bottom
- Slide sling under client
- Repeat the process for the other side of the client's body

Chair to Chair

1

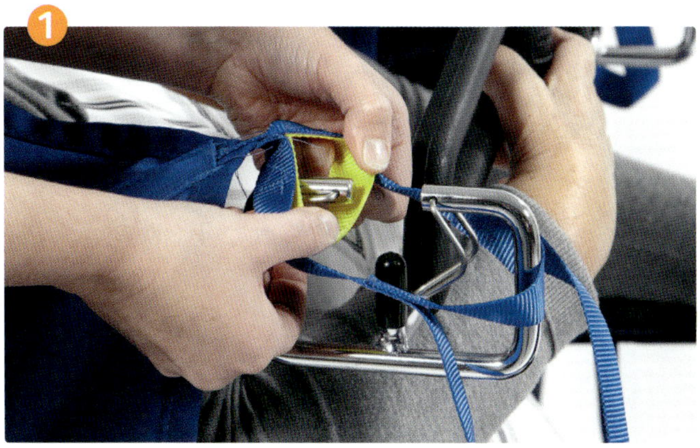

- Follow manufacturer's instructions for the specific hoist
- Having fitted a sling, connect sling straps to hoist
- Leg straps can be interlocked or crossed to prevent client's legs from spreading to protect their dignity

NOTE:
If a choice of strap length is available it is important to have two the same length each side for the shoulders, the pelvis and legs. Shoulder straps may be a different length to leg straps. This depends upon where the client is being moved to or from, e.g. chair to chair would require short straps.

2

- Explain to client what is going to happen to reassure them
- Complete a final check of attachments before hoisting begins
- Leavc brakes **off** during hoisting as hoist will need to adjust to its best balance position

3

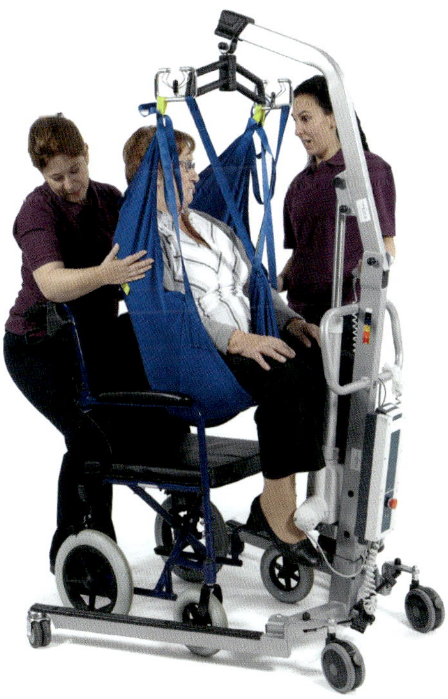

- Using hoist, lift the client enough to clear the chair
- Using hoist, move the client to new location
- One carer should remain where the client can see them for reassurance

NOTE:
Keep travelling distances to a minimum. Do not suspend client in the air for long distances.

4

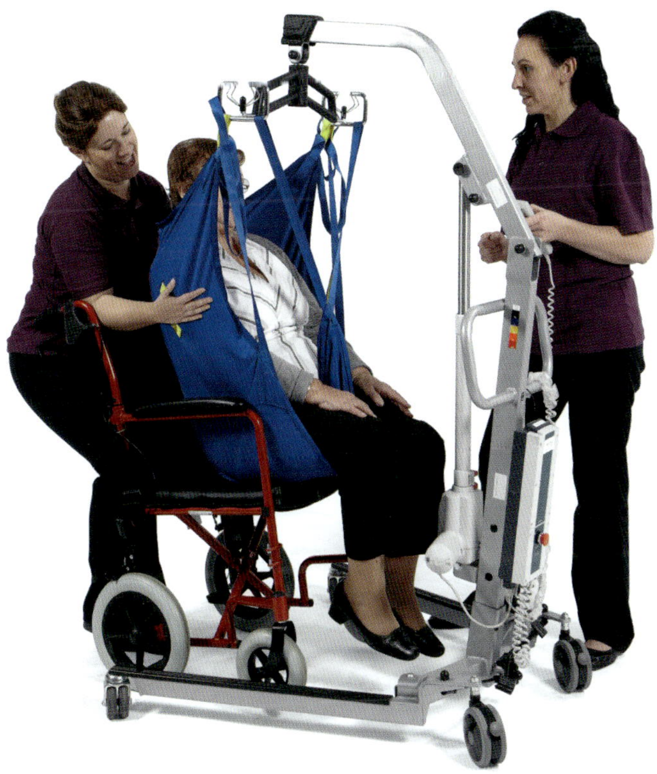

- Once positioned over the chair, lower the client using hoist
- Guide the client back onto the chair

Fitting a Sling to a Client in Bed

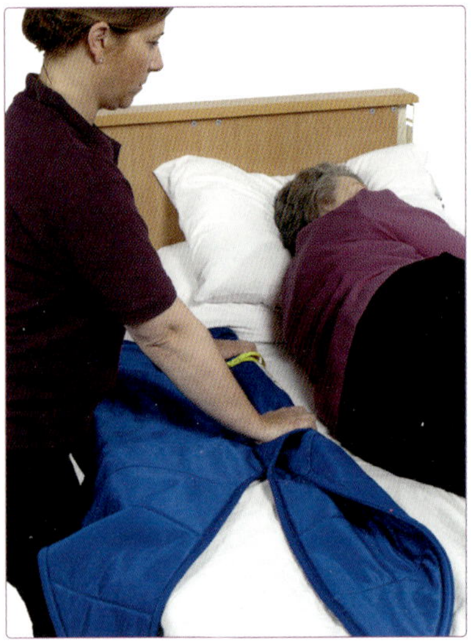

- Roll client onto their side
- Roll one half of the sling, with insides touching, to the centre of the sling

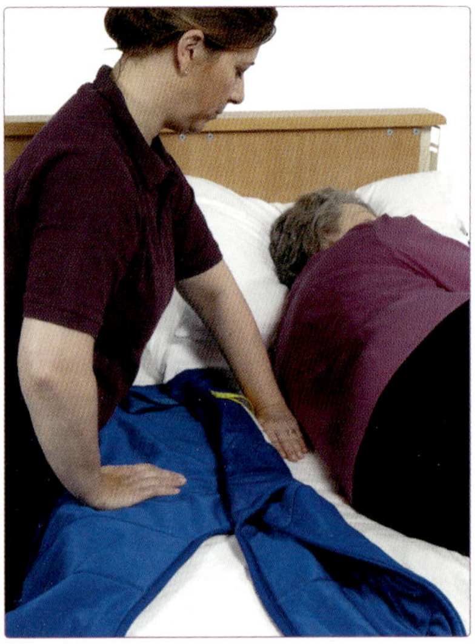

- Place the rolled edge of the sling next to client's spine with a hand's width between the edge and the spine

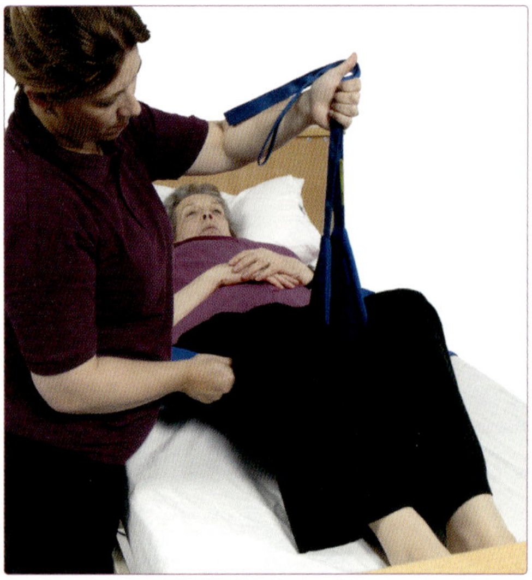

- Roll the client onto the sling and tilt them slightly towards yourself so the rolled sling section can be unrolled under client
- Feed the leg straps between the client's legs
- Make sure strap lengths are even on both sides before connecting to a hoist

CONSIDER
Could the bed height be adjusted to make movement easier?

From the Bed

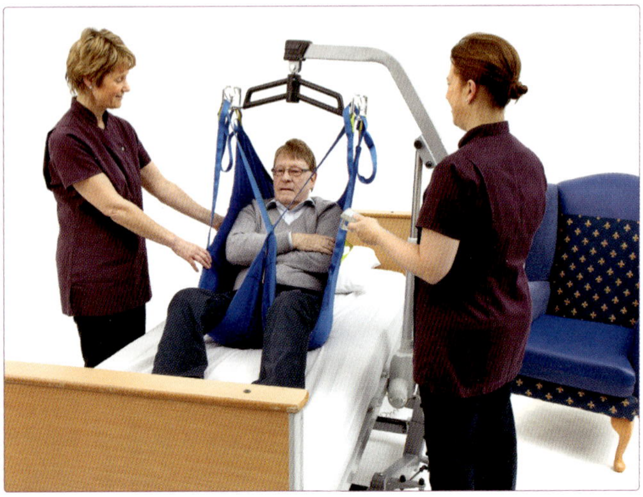

- Follow manufacturer's instructions for the specific hoist
- Having fitted a sling, connect sling straps to hoist
- Using sling, hoist client from bed
- Once client is clear of the bed move the hoist away from the bed

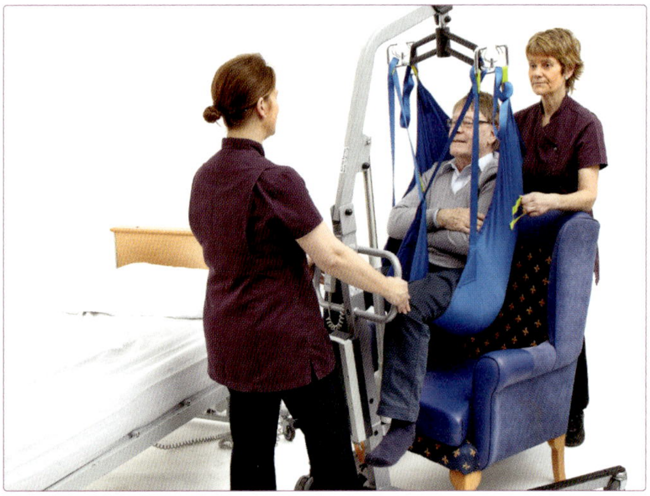

- Client may need to be rotated on the boom arm before repositioning on the chair
- Turn the sling bar and carefully position the client's legs so that they are on either side of the hoist's mast
- Move hoist closer to the chair
- Once in the correct position, lower client onto chair

NOTE:
Keep travelling distances to a minimum. Do not suspend the client in the air for long distances.

From the Floor

- Follow manufacturer's instructions for the specific hoist
- Having fitted a sling, one carer should kneel behind the client
- Other carer should approach client from the side with the hoist
- Create two gaps under client for the legs of the hoist *(i.e. under the client's knees and behind their back)* by sitting the client slightly forward and supporting with a carer or chair
- Connect sling straps to hoist

NOTE:
Keep travelling distances to a minimum. Do not suspend the client in the air for long distances.

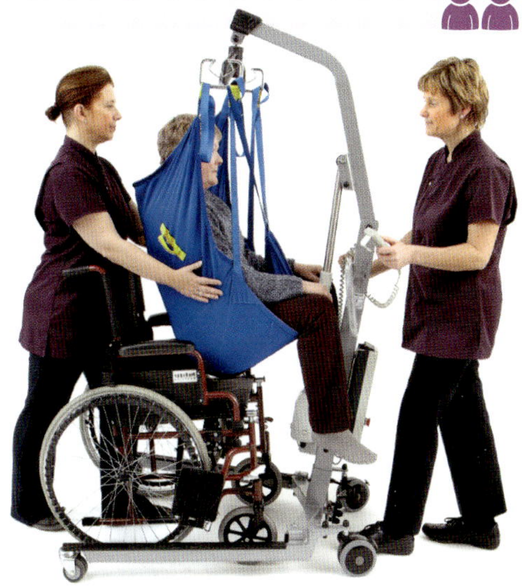

- Hoist client from the floor
- Client may need to be rotated on the boom arm before repositioning on the chair
- Turn the sling bar and carefully position the client's legs so that they are on either side of the hoist's mast
- Use hoist to move client to a bed or chair
- Once in the correct position, lower client onto bed or chair

Standing Hoist

or

There are occasions where full hoisting may not be required. If hoisting from a sitting position and the client is able to weight-bear, a standing hoist may be more appropriate.

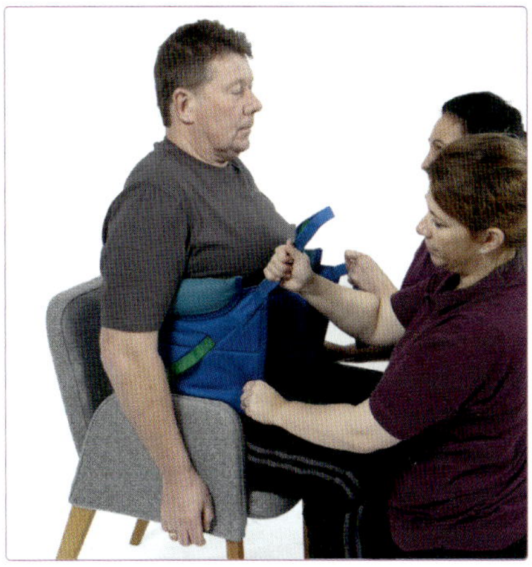

- Follow manufacturer's instructions
- Fit specified sling around the client's back
- Make sure sling fits snugly and the padded area is against the chest wall and NOT directly under the client's shoulder joint

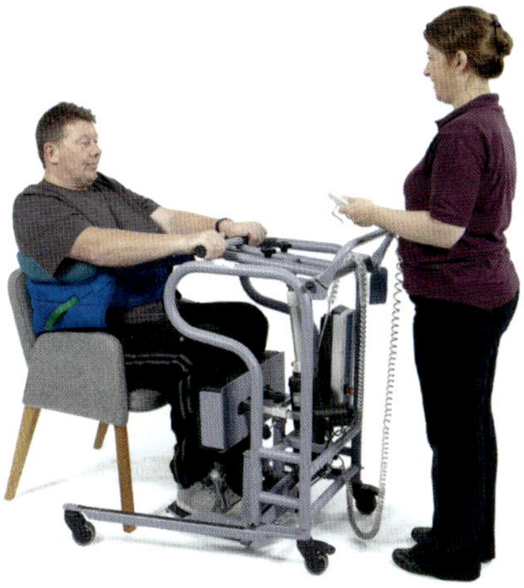

- Position knee supports on standing hoist just lower than the client's knee joints
- Make sure the client's feet are on hoist footplates and they are sitting towards the back of the chair
- Put the brakes on
- Connect sling to the hoist
- Ask client to lean back against sling and hold the hoist handles

- Take the brakes off if recommended in the manufacturer's instructions
- Hoist client off the surface they were sitting on
- It may not be necessary to bring them up into a full standing position

NOTE:
This is an ideal aid for toileting.

Fallen Client

Knowing what to do when a client has fallen is important. A risk assessment of a client's potential to fall should be included in the client's personal handling plan *(see page 19)*.

CLIENT HAS FALLEN

Injured

Call for an ambulance and explain suspected injury

No apparent injury

Reassure the client

Provide comfort and support

DO NOT RUSH THE CLIENT TO GET UP

↓

Encourage the client to get up independently *(see next page)*

Use verbal prompts

Bring secure furniture near to help client get themselves off the floor

↓

If client is unable to get up use equipment, e.g. hoist or emergency lifting cushion

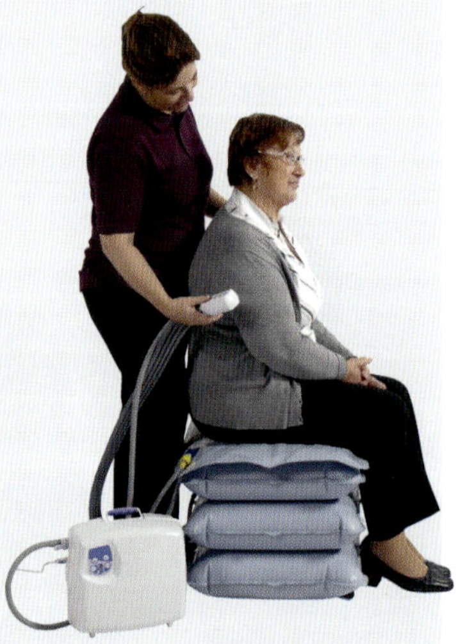

If this equipment is not available an ambulance or falls service may be required

↓

Once client has been transferred to a bed or chair, if you are still concerned about any potential injury contact a medical practitioner, e.g. GP

CAUTION
If a client has fallen they should not be lifted manually from the floor.

Instructing a Client to Get Up from the Floor with Minimal Assistance

Before instructing a client to get up from the floor, you should make sure that they are not injured. The client should be given the information stage by stage and encouraged to take their time.

1

- Ask client to bend their knees, place one arm away from their body and bring their other arm across their chest

2

- Ask client to roll towards their outstretched arm onto their side and bring their other arm over their body until their hand is flat on the floor

3

- Ask client to push on their forearm and other hand until their upper body and shoulders have risen off the floor

4

- Encourage the client to keep pushing until they are on all fours in a crawling position

5

- Position a chair at the head of the client
- Ask the client to put their hands and forearms on the seat of the chair
- Encourage client to lean on the seat of the chair, lift one knee up and place their foot flat on the floor

6

- Ask the client to push up to straighten their legs

7

- Encourage client to turn and sit on the seat

Emergency Handling

It is vital to have a plan for moving clients in an emergency situation should they be unable to move themselves.

Chair to Floor in an Emergency

The Resuscitation Council *(UK)* recommends that carers should not attempt to lift a client onto a bed, but should instead slide the client on to the floor to perform emergency techniques such as CPR. The main aim is to slide the client from the chair while making sure their head is safe.

1

- Call for help
- Kneel next to the chair using a wide base of support
- Maintain your spinal curves
- Place client's nearest arm across their chest

2

- Push client's knees away from yourself
- Slide arm behind the client and hold their furthest hip

3

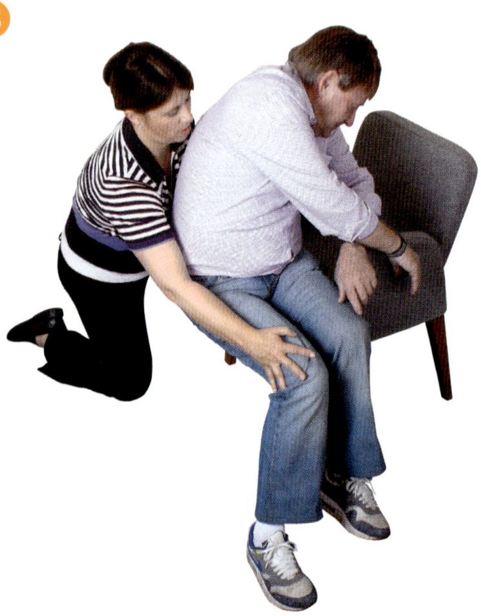

- Slide client's hip forward to the edge of the seat
- Continue pressure on client's knees and hip

4

- Once their bottom is over the edge of the seat, move support from client's legs to their head. Client will lower to the floor with gravity
- Continue to support the client's head as they lower to the ground

Chair to Floor in an Emergency

1

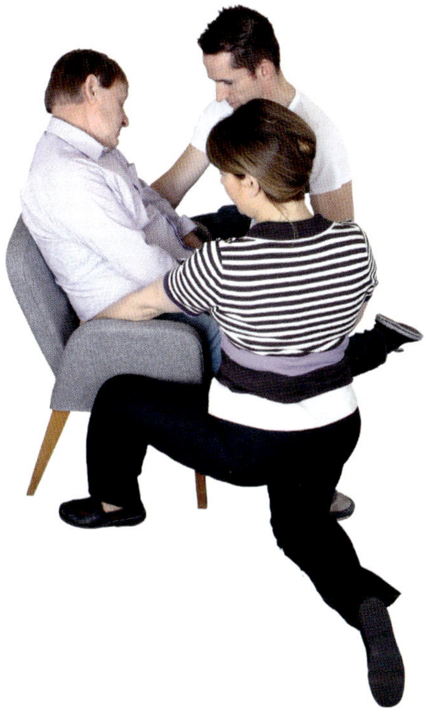

- One carer should kneel on either side of the chair facing the client
- Both carers should adopt a high kneeling position on one knee
- Maintain your spinal curves

2

- Place outside hand behind client's pelvis
- Place inside hand around inside of client's knee

3

- Use command such as "Ready, steady, slide"
- Carers should sit back on their heels and slide the client's pelvis forward in the chair

4

- One carer should support client's head
- One carer should look after client's legs
- Lower client into a flat position on the floor

Handling a Client That has Fallen in a Confined Space

Clients often collapse in small confined areas, e.g. toilets, where it is not possible to use lifting equipment such as hoists. In these cases special slide sheets can be used to move the client from the confined space into a larger area where a hoist can be used.

Unconscious Client

1

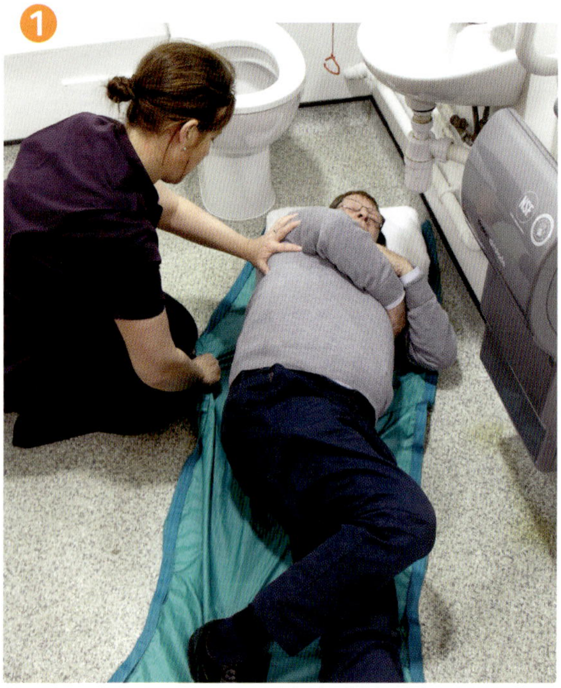

- Cushion client's head, e.g. use a pillow or towel
- Place one full length slide sheet under the client with the non-slip side under the client *(see page 43)*
- Attach two straps to the end of the sheet

2

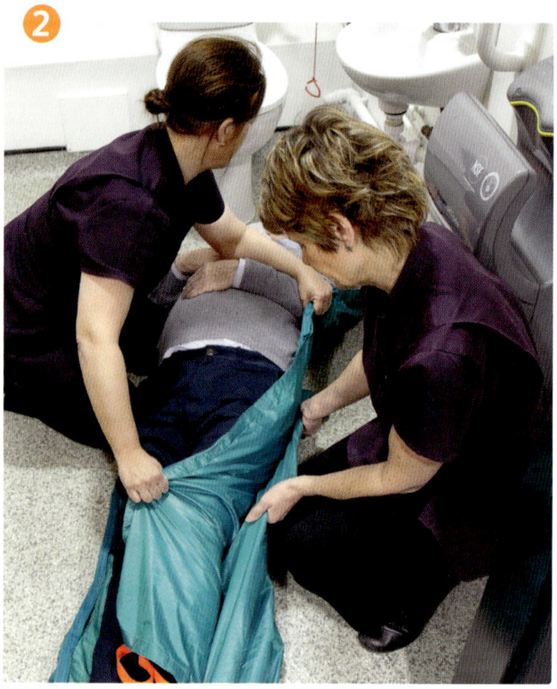

- Place a second slide sheet, slippery side up, half way under the first slide sheet. The excess should be in the direction you want to move the client.

3

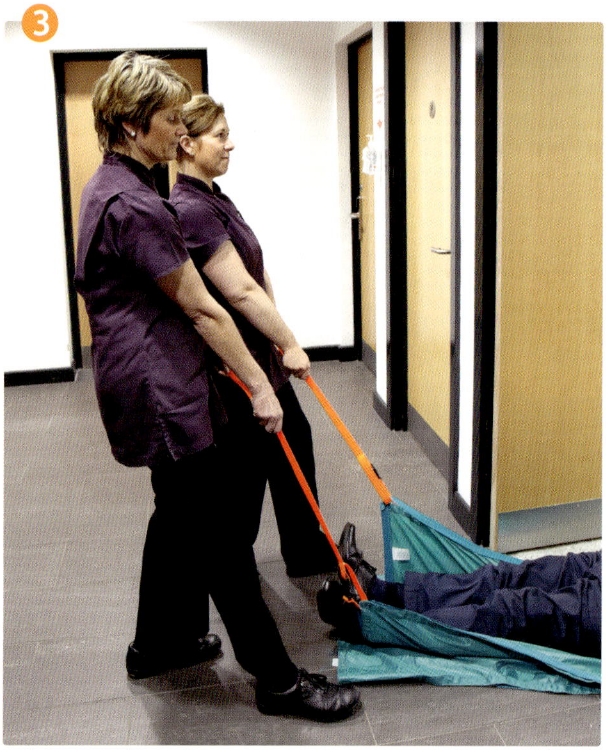

- Carers should stand holding one strap each
- Adopt a safe handling posture
- Use command such as "Ready, steady, walk"
- Walk backwards in the direction you want to move the client
- Reposition the bottom slide sheet as required

Handling a Client That has Fallen in a Confined Space

Conscious Client

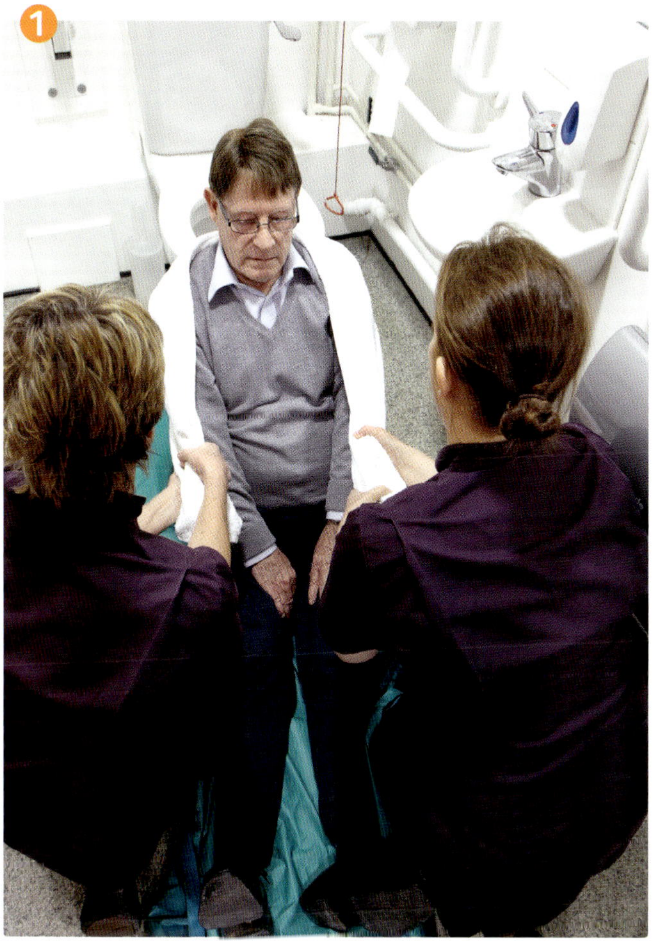

- Place one full length slide sheet under the client *(slippery side under the client)*
- Carers should kneel either side of the client
- Adopt a safe handling posture
- Place a towel around the client's shoulders *(avoiding the neck)*
- Carers should hold the towel taut

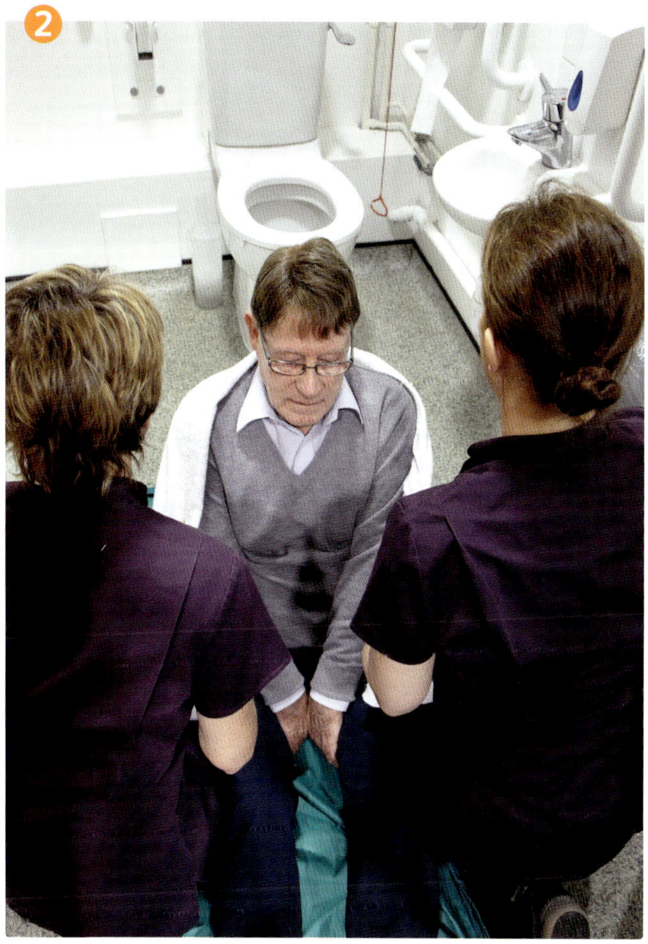

- Use command such as "Ready, steady, slide"
- Use the taut towel to pull the client towards you

CONSIDER

Check there are no signs of wear on the towel before use.

Transferring a Client Down Stairs

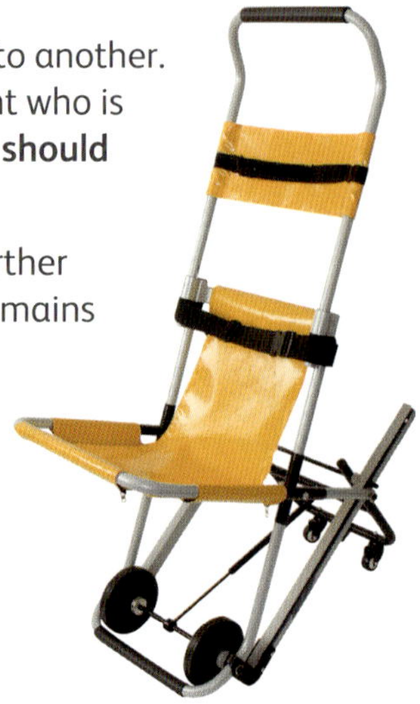

Whenever possible, lifts should always be used to transfer clients from one floor level to another. If lifts are not available, e.g. during a fire, an alternative method of evacuating a client who is unable to move themselves should be planned for and risk assessed. **Special training should always be given before using equipment to transfer clients down stairs.**

For bed-bound clients, straps are available that can secure a client to a mattress. Further straps then attach to the mattress to pull the mattress down the stairs. The client remains cushioned by the mattress and the carers are able to slide and not lift the client.

Evacuation Chair

For a seated client, an evacuation chair can be used. This chair-shaped device is more suitable to move a client down stairs than a wheelchair or carrying the client.

Moving a Plus Sized Person

As a population we are increasing in weight, so management of the plus sized person has become more of a consideration. The body shape of the individual can affect how they can perform certain tasks. Careful consideration should be given to assessing each individual client using their personal handling plan *(see page 19)*.

Use of equipment can significantly reduce the risks for both the carers and the client and can help to improve the client's function. Care should be taken to make sure any equipment used has a suitable weight limit. Most equipment will have a weight restriction but versions are available for the plus sized person *(see pages 32–34)*. Sometimes the simplest equipment *(such as slide sheets)* can create the most benefit.

Using a Wheelchair

The principles of safer handling when pushing and pulling loads are outlined on page 23 and if applied will help to protect the carer. However when the load is a person, there are other areas that need to be considered.

Before transferring a client to or from a wheelchair:

- Apply the brakes
- Put the footrests in an upright position and out of the way

Before pushing a client in a wheelchair make sure:

- Footrests are in the flat position
- Client's feet are on the footrests
- Client's arms and legs are suitably placed
- Safety belts are used as appropriate
- Covers are used as appropriate to protect client's dignity

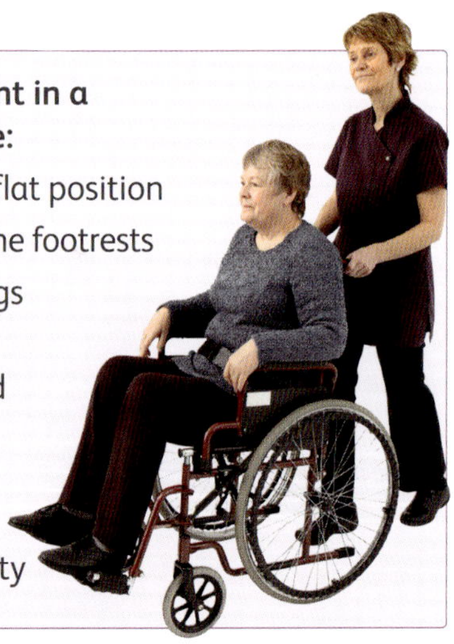

When pushing a client in a wheelchair:

- Aim for a smooth ride
- Do not lean down on the wheelchair handles *(this may tip the client unnecessarily)*
- Avoid uneven surfaces such as cobbles
- Go over unavoidable bumps slowly
- Use a zigzag pathway rather than a straight line when steering downhill to give you more control
- Avoid using wheelchairs in certain weather conditions, e.g. snow and ice

Kerbs

Use ramps wherever possible and exit a pavement using a dropped kerb.

Negotiating a wheelchair up a kerb

If it is safe to do so:

- Approach the kerb head on
- Position the front wheels so they nearly touch the kerb
- Tell the client you are about to tip the wheelchair
- Hold the handles firmly and push downwards and use your foot on the tipping lever to push downwards
- Tip the wheelchair so it is balanced on its back wheels
- Push the wheelchair forwards so that front wheels are over the pavement and the back wheels are touching the kerb
- Lower the front wheels gently onto the pavement
- Push the back wheels onto the pavement keeping in contact with the kerb at all times
- **NEVER** lift the wheelchair onto the kerb
- **NEVER** lift the back wheels off the ground

Negotiating a wheelchair down a kerb

If it is safe to do so:

- Approach the edge of the kerb head on
- Tell the client you are about to tip the wheelchair
- Hold the handles firmly and push downwards and use your foot on the tipping lever to push downwards
- Tip the wheelchair so it is balanced on its back wheels
- Lower the wheelchair slowly down the kerb making sure both back wheels touch down at the same time
- Gently lower front wheels back onto the ground

Taking the wheelchair down backwards may be easier for plus sized clients but has the extra danger of the carer having their back to the traffic and having to make a full turn in the road before they can push the wheelchair any further.

NOTE:
Wheelchairs should be regularly inspected, well maintained and carers trained in their use, as required by the Provision and Use of Work Equipment Regulations *(1998) – (PUWER)* *(see pages 27 & 34).*

Tips for Back Pain

Changing your activities as described in this book should help reduce the risk of back injury. However, if you already experience back pain, here are some other things to consider:

Keep active

Even when you have pain, regular movement has been shown to reduce pain compared with long periods of bed rest.

Exercise regularly

Exercises such as walking and swimming will help to strengthen your back muscles and activities such as yoga and pilates are particularly designed to improve posture and flexibility.

Healthy diet and lifestyle

Healthy diet and lifestyle has been shown to improve back pain. Studies show that factors such as smoking can increase the risk of back injury and developing back pain.

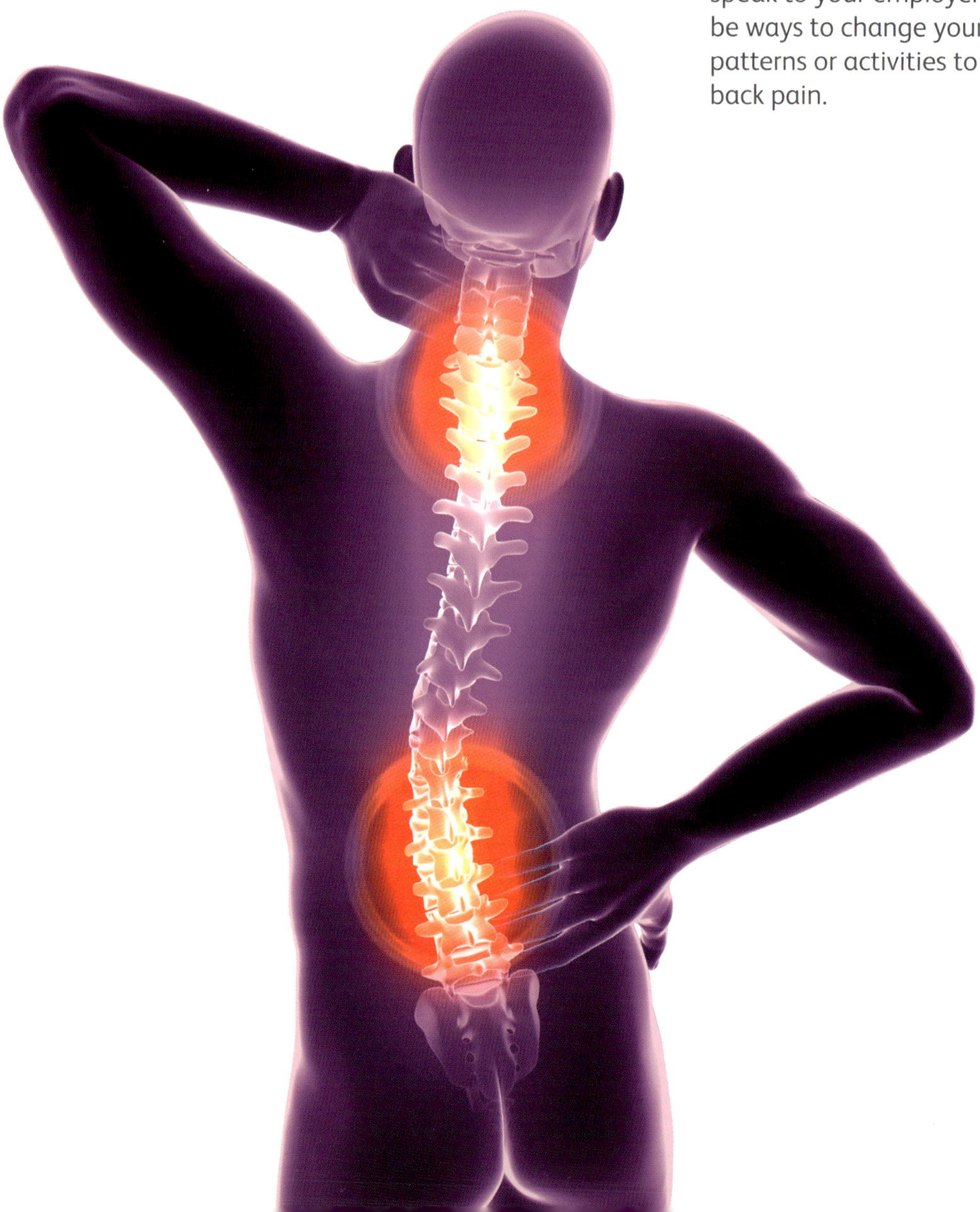

Pain relief

Painkillers such as paracetamol, ibuprofen, or pain relieving gels can help you to maintain activity. If the pain does not improve, seek further medical advice from a GP or health professional.

Carrying

Avoid carrying bags on one side. Consider across shoulder or rucksack type bags.

Think tall

Slouch postures will encourage back pain. Good posture reduces spinal pressures and makes you look taller.

Work environment

If the nature of your job role can not be easily adapted, speak to your employer or HR department. There may be ways to change your work environment, work patterns or activities to enable you to manage your back pain.

Test Your Knowledge

1. **What is the safest position in which to carry an object?** *Page 20*
   A. Close to the floor.
   B. Close to your body.
   C. At arms length.
   D. Over your shoulder.

2. **What does T.I.L.E. stand for in Risk Assessment?** *Page 18*
   A. Team, Individual, Leader, Execute.
   B. Tilt, Incline, Lean, Enable.
   C. Task, Individual, Load, Environment.
   D. Tools, Industry, Levers, Engage.

3. **What is it most important to do when lifting as a team?** *Page 24*
   A. Appoint a leader and communicate well.
   B. Work in friendship groups.
   C. Have good hygiene procedures.
   D. Take turns holding the heavy end.

4. **According to Health and Safety and Manual Handling Laws, what should employees make their employer aware of?** *Page 7*
   A. Danger in the workplace.
   B. New technology.
   C. Unauthorised absences.
   D. Feeling undervalued.

5. **When risk assessing a load which of the following aspects is the most important?** *Page 17/18*
   A. Colour.
   B. Weight.
   C. Branding.
   D. Value.

6. **Which of the following is key information that should be included in a client's personal handling plan?** *Page 19*
   A. Blood type.
   B. Cholesterol level.
   C. Level of assistance needed.
   D. Nationality and religion.

7. **What is the first thing you should do if a client has fallen?** *Page 50*
   A. Get them up as quickly as possible.
   B. Find out if they are injured or not.
   C. Lift them off the floor.
   D. Call an ambulance immediately.

8. **Which of the following is an advantage of using a handling belt?** *Page 32*
   A. Can be used to lift a client up from the floor.
   B. Applies firm pressure to the client's clothes and skin.
   C. One size is suitable for everybody.
   D. Provides a good handling grip and preserves the client's dignity.

Notes

Notes

Notes

Notes

Glossary

Accident
An unplanned or unexpected event that causes damage to persons, equipment, stock, buildings etc.

Accident book
Forms on to which details of all accidents may be recorded.

Arthritis
Inflammation of a joint that can be caused by wear and tear due to repetitive tasks.

Breach
The infringement of a legal right or duty.

Cartilage
Firm rubbery tissue that cushions bones at the joints and allows for smooth movement.

Centre of gravity
The point at which the whole weight of an object can be considered to act and therefore at which all parts of an object are balanced.

Competent
Able, skilful and properly qualified.

Competent person
A person who has sufficient training, experience, knowledge or other relevant qualities to enable them to undertake a task/job.

Consent
To give approval or permission. A person must be of sufficient mental capacity and of the age at which they are legally recognised as competent to give consent *(age of consent)*.

Control measure
Action or procedure put in place to reduce risk to an acceptable level.

CPR
Cardiopulmonary Resuscitation. Manually squeezing the heart and breathing for a patient.

CQC
Care Quality Commission. Regulator of health and social care in England.

Cross contamination
A common term for the way harmful bacteria pass from one object to another.

Discs
A disc shaped piece of specialised tissue that separate the vertebral bones of the spinal column. They enable the spine to flex and extend, allow smooth movement of the spine and help bond the vertebrae together.

Ergonomics
The application of scientific information concerning human beings and the design of objects, systems and environments for human use.

Force
A push or pull on an object resulting from two objects interacting with each other.

Fracture
A break in bone or cartilage.

Fulcrum
The point or support on which a lever rests or pivots.

Hazard
Something that has the potential to cause harm.

Hernia
A protrusion of tissue through a weakness in the muscle or surrounding tissue wall.

Hierarchy of control
Control measures listed in order of priority.

HSE
Health and Safety Executive. Non-departmental public body responsible for workplace health, safety and welfare.

Liability
Legal obligation or duty.

Ligaments
Band of fibrous tissue connecting bones or cartilage. Can hold joints together.

Load
An object to be moved, held or positioned.

LOLER *(1998)*
Lifting Operations and Lifting Equipment Regulations *(1998)*. Regulation for lifting equipment at work.

Manual handling
Using physical effort to lift, carry, push or pull a load.

Musculoskeletal disorders
Injury to the musculoskeletal system including problems such as lower back pain, joint injuries and repetitive strain injuries of various sorts.

Musculoskeletal system
The system in the body of bones, muscles, ligaments and tendons.

Nucleus
Inner gelatine like core of the spinal discs.

Occupational health
The promotion and maintenance of physical and mental well being of all staff.

Personal handling plan
Documentary evidence of the recommended handling scheme, treatment and care of an individual.

Personal Protective Equipment *(PPE)*
Personal Protective Equipment or equipment worn to protect the wearer against hazards.

PUWER *(1998)*
Provision and Use of Work Equipment Regulations *(1998)*. Regulation for any equipment used at work.

Reasonably practicable
A risk/sacrifice computation made to see what safety measures are reasonable under given circumstances of cost/benefit.

RIDDOR *(2013)*
Reporting of Injuries, Diseases and Dangerous Occurrences Regulations *(2013)*. Regulation that requires certain injuries, diseases and incidents at work to be reported to the HSE.

Risk
The likelihood of the hazard actually doing damage.

Risk assessment
A proactive technique for preventing accidents and illness by identifying problems and introducing control measures.

Slipped disc
When the layers of cartilage of the disc become cracked and the inner nucleus pushes out of the crack and presses on the spinal nerve.

Spinal cord
The major column of nerve tissue that is connected to the brain and lies within the spinal canal.

Spinal nerves
Nerves that originate in the spinal cord. There are 31 pairs of spinal nerves.

Sprain or strain
The muscle or ligament fibres tear as a result of overstretching.

Stability
The ability of an object to return to a balanced position with the same base of support after being disturbed.

Static work
Any posture held for longer than 30 seconds.

Tendon
Fibrous band of connective tissue by which muscles attach to bones.

T.I.L.E.
Task, Individual, Load, Environment. Considerations when risk assessing a manual handling task.

Vertebrae
Plural of vertebra. A vertebra is one of the 33 bony segments that form the spinal column of humans.

Vibration
Continuous quick, slight shaking movement.

Young person
A person who has ceased to be a child *(over school leaving age)* but is not yet 18.